Avoiding Disaster Deaths

Nibedita S. Ray-Bennett

Avoiding Disaster Deaths

Why Do So Many People Die?

Springer

Nibedita S. Ray-Bennett
School of Business, Avoidable Deaths
Network, Institute for Environmental
Futures
University of Leicester
Leicester, UK

ISBN 978-3-031-81299-6 ISBN 978-3-031-81300-9 (eBook)
https://doi.org/10.1007/978-3-031-81300-9

This Springer imprint is published by the registered company Springer Nature Switzerland AG
The registered company address is: Gewerbestrasse 11, 6330 Cham, Switzerland

To my late parents, Mr. Kiran Shankar Ray and Mrs. Sarada Ray

To those who lost their loved ones during the COVID-19 pandemic

Acknowledgements

This book is based on the work of many brilliant minds in the Avoidable Deaths Network (ADN), and I am heavily indebted to them for their contributions.

First, our ADN content review team: Dr. Hideyuki Shiroshita, Dr. Vincent Pieterse, Dr. Azukaego Nnaji, Dr. Edris Alam, Dr. Madhulika Sahoo, Dr. Bhola Saha, Dr. Srastha Chowdhury, Dr. Biswajit Paul, Dr. Lopamudra Saxena, Mr. Daniel Mendez, Ms. Shaiza Khwaja, Ms. Kumiko Fukuromoto, Mr. Krishna Clarke, Ms. Anuradha Bhardwaj, Ms. Rwittambhara Kashyap, Mr. Lakshman Srikanth, Mr. Tanmay Gound, Ms. Aarju Basnet, Mr. Aryabhusan Goutam, Ms. Nimisha Goswami, and Mr. David Cowie. I thank Mr. Alex Skinner and Mr. Bede Wilson for sharing the burden of proofreading and Mr. Julian Coetzee and Mr. Lasith Dissanayake for assisting me with the references for each chapter.

I would like to thank Mr. Dillip Pattanaik for funding the newspaper subscriptions. Thanks are also due to the University of Colorado's Hazard Research Centre for the One-Off award for this study under the remit of 'COVID-19 Working Groups for Public Health and Social Science Research'.

Special thanks are due to my several research assistants who helped immensely in different phases of this study: Ms. Antara Ragini Dev Sarmah, Ms. Lauren MacLeod, Mr. Lasith Dissanayake, and Dr. Denise Corsel. I am also grateful to our project's physician, Dr. Azukaego Nnaji, for assisting with finalising the content review of several causes of death and drafting of the preventable and amenable avoidable death indicators.

I greatly thank our ADN advisors, Dr. Ambika Prasad Nanda and Mr. Mihir Bhatt. This book would not have come to fruition without their encouragement to start this project at the height of COVID-19. I am grateful to Prof. Christina Hughes for helping me with the title of this book and Prof. Norio Okada and Prof. Michael Petterson for helping me clarify my thinking around the avoidable disaster deaths approach. I thank Mr. Steve Glovinsky for his support and advice in implementing this approach through India's Avoidable Snakebite Deaths Project. Finally, I would also like to thank my longstanding mentor and advisor, Prof. Heiko Balzter, for his continued support and encouragement in all my research endeavours for over a decade.

I thank all the reviewers for providing critical feedback to improve the book's quality.

I thank my dearest late parents, Mr. Kiran Shankar Ray and Ms. Sarada Ray, who gave me the freedom to be educated, marry the love of my life, and live a full and accomplished life every day. I also thank my dearest husband, Adam Bennett, and our daughter, Asha Ray Bennett, for their patience and support throughout the writing of this book. I hope Asha reads this book when the time is right and learns more about mummy's work.

We have lost far too many lives in the COVID-19 pandemic. Whilst writing this book over four years, their passing has been a constant reminder. As an author, I hope to bring closure to the grieving and move forward.

About This Book

Disaster deaths are the direct and indirect impact of natural, technological, or biological hazards. The 'Sendai Framework for Disaster Risk Reduction' has urged all United Nations Member States to reduce disaster deaths by 2030. Reducing the number of disaster deaths has become a valuable statistic for improving disaster risk management. Equally important is to know how people die, who dies, why they die, what they are dying from, and which disaster deaths are 'avoidable' and 'unavoidable'. Disaster risk reduction specialists have not fully examined these questions.

Avoiding Disaster Deaths introduces a novel approach called 'Avoidable Disaster Deaths' (ADD) to explore these questions. Built from disaster deaths, public health and epidemiology, human geography, risk and crisis management studies, this approach disaggregates disaster deaths as avoidable and unavoidable. Avoidable disaster deaths are preventable, amenable, and governance-related. Unavoidable deaths occur annually due to natural causes and poverty-related diseases. In a disaster climate, the emphasis should be on reducing avoidable disaster deaths in a resource-constrained context to reduce excess deaths.

The analytical advantage of this approach is explored using the context of the first wave of COVID-19 and the economic lockdown in India. The book presents evidence of risk governance adopted to implement lockdowns and unlocks and real cases of avoidable disaster deaths. It introduces novel matrices and dynamic strategies to ascertain the 'causes and circumstances' of avoidable disaster deaths for small-scale prospective projects.

The book is suitable for students, academics, policy-makers, and practitioners interested in disaster risk reduction, human rights, risk and crisis management, environmental science, human geography, Sustainable Development and Sendai Goals.

The book is also suitable for passionate citizens who want to capture the number, causes, and circumstances of avoidable disaster deaths and take positive action to save lives in their communities.

Contents

About the Author

Nibedita S. Ray-Bennett is Professor of Risk Management at the University of Leicester School of Business. She is one of the founding and Presidents of the Avoidable Deaths Network (ADN) and Associate Director at the Leicester Institute for Environmental Futures.

Nibedita has Ph.D. in Sociology from the University of Warwick, which was funded by the Ford Foundation's International Fellowship Programme. Nibedita is Fellow of the Higher Education Academy Fellow of the Royal Geographical Society and Global Expert at the International Scince Council.

Nibedita's research is inter-disciplinary and transdisciplinary and it focuses primarily on reducing avoidable disaster deaths in lower-middle and low-income countries through novel theories, methods, interventions, lifesaving kits, networking, public engagement, and advocacy.

In collaboration with Kansai University and the city of Izumiotsu in Osaka, she launched the global campaign 'International Awareness Day for Avoidable Deaths' (IAD4AD) in 2019, with the campaign slogan 'Disaster Deaths Are Avoidable'. This campaign aims to raise the visibility of indirect disaster deaths and missing persons and to understand the causes and circumstances surrounding avoidable disaster deaths.

Nibedita is the author of *Caste, Class, and Gender in Multiple Disasters* (VDM Verlag, 2009) and *Avoidable Deaths: Systems Failure Approach to Disaster Risk Management* (Springer Nature, 2018).

List of Figures

List of Tables

Chapter 1
What is a Disaster?

> *[…] as our systems of response become more and more sophisticated for recurring disasters, and people take heed of familiar threats, the disaster can still be challenging and lethal. How people die matters.*
> —Kayyem (2022)

Abstract The number of disaster deaths is a valuable statistic, but to improve disaster response, it is essential to know how people die, why they die and which deaths are avoidable. To make sense of disaster deaths, this chapter introduces a novel approach, 'avoidable disaster deaths' (ADD), which disaggregates or breaks down deaths into avoidable and unavoidable in a disaster climate. Avoidable disaster deaths are preventable, amenable, and governance-related. Unavoidable deaths are those caused by natural causes, poverty-related diseases, or other causes, annually. By engaging with mainstream disaster risk reduction theory, this chapter redefines COVID-19 as a 'biological disaster' to capture the number of avoidable disaster deaths in the first phase. This chapter also introduces the case study of COVID-19 in India and the data-gathering procedures undertaken from March to December 2020. The novelty and limitations of the data are also discussed.

Keywords Avoidable disaster deaths · Biological disaster · COVID-19 · Curbing the virus · Economic lockdown · Disaster deaths · India · Secondary data

Introduction

While writing this book, Professor David Spiegelhalter and Dr. Anthony Masters published *Covid by Numbers* (2021). They wrote something compelling in the opening of their Introduction chapter: "Life changed in March 2020, and everyone has their story to tell" (Spiegelhalter & Masters, 2021, p. 1). Indeed, our lives changed forever. Indeed, everyone has a story to tell. And yet, some stories need to be told on behalf of the deceased's family members.

N. S. Ray-Bennett, *Avoiding Disaster Deaths*,
https://doi.org/10.1007/978-3-031-81300-9_1

According to Kayyem (2022), "how people die matters" and "we must accurately memorialize how people die" so that lessons can be learned to avoid disaster deaths. Also, "what are people dying from?—so that they guide decisions in public health and find ways to save lives" (Dattani et al., 2023). This book aims to do so by foregrounding the "circumstances and causes" that led to avoidable disaster deaths in India's first phase of COVID-19. Unfortunately, the burden of these deaths was borne mainly by the socio-economically vulnerable sections of society. These deaths could have been avoided through foresight, reflective response, and effective disaster risk governance—among other things.

I have been working in the domain of disaster risk reduction (DRR) and international development for over two decades. My personal experience as a relief worker in the aftermath of the Super-Cyclone in 1999 in Odisha, India, brought me to this world of disaster reduction and avoidable disaster deaths. In 1999, I was reading my Master of Arts in Social Work at the Tata Institute of Social Sciences in Mumbai. On 17–18 October 1999, Odisha, an eastern State of India, was affected by a category 4 tropical cyclone (wind velocity of 270–300 km per hour), which killed more than 10,000 people. After the Super-Cyclone, I went to Odisha to work as a relief worker through the Tata Institute of Social Sciences. Under the guidance of the District Disaster Management Officer, I spent a fortnight in the district of Jagatsinghpur, which bore 70% of disaster deaths. I assisted the local government in caring for widows, orphans, and the elderly impacted by the Super-Cyclone. This experience was instrumental in shaping my life and career. Based on this experience, I was able to secure the Ford Foundation's International Fellowship Programme to conduct my PhD research on 'Caste Class and Gender in Multiple Disasters: Women-Headed Households Experiences in Odisha' in the Department of Sociology at Warwick University in the United Kingdom. During my fieldwork for my PhD research in 2003, I was reunited with the women I had assisted in the aftermath of the Super Cyclone. I was also able to document their experiences in depth, not only of the Super Cyclone but also of later disasters, which included the floods in 2001 and 2003 and the drought in 2002 (Ray, 2006; Ray-Bennett, 2009).

After a decade, I returned to Odisha in 2013, this time to study another tropical cyclone called Phailin. On 12 October 2013, Odisha was hit by another category 4 tropical cyclone (wind velocity 260 km/h). Only 86 people died. I wanted to study how the Government of Odisha drastically reduced the number of disaster deaths from 10,000 in 1999 to 86 in 2014. I deciphered the reasons for lowering disaster deaths through the perspective of the systems failure approach, which led to my book *Avoidable Deaths: Systems Failure Approach in Disaster Risk Management* (Ray-Bennett, 2018). This book also led to the launch of the Avoidable Deaths Network[1] (ADN) in 2019. ADN is a global–local network of more than 900 members in 94 countries dedicated to finding theoretical and practical solutions to reducing disaster deaths and injuries for sustainable development in lower-middle—and low-income countries.

---

[1] https://www.avoidable-deaths.net/.

In March 2020, I planned to launch the ADN India Hub in Odisha, but this was cancelled due to the COVID-19 pandemic. Instead, I stayed at home in England and watched and heard the horror of deaths unfolding on my television, through print media, and word of mouth from my friends, family, and colleagues back in India. I never felt so helpless in my life. An active and direct response was impossible.

After much deliberation with the host of ADN India Chapter and Advisors, I decided to respond indirectly through research by collecting secondary data on avoidable disaster deaths that were taking place from the impact of COVID-19 and the economic lockdown. This begged the question: how can I do this? How can one make sense of which deaths are avoidable and which are unavoidable in a disaster climate? And how can I capture them remotely? These led to the conception of two central research questions for this book:

- How many avoidable disaster deaths were occurring due to the impact of COVID-19 and the economic lockdown in India, and how can we capture them?
- What are the circumstances and causes surrounding these deaths?

To capture the number of avoidable disaster deaths, I conceived a novel approach, '**avoidable disaster deaths**' **(ADD)**. The 'avoidable disaster deaths' approach is distinct from the 'avoidable death' and 'disaster deaths' approaches. The former is used amongst public health and epidemiology experts and the latter amongst the advocates of the United Nations Office for Disaster Risk Reduction's (UNDRR) 'The Sendai Framework for Disaster Risk Reduction 2015–2030' (hereafter, the Sendai Framework). I have resumed the discussion on these in Chaps. 2 and 3.

Avoidable disaster deaths are understood as premature preventable, amenable, and governance-related deaths. These deaths are avoidable through tailored, preventable, amenable (Cook, 2019; Price, 2021) and risk governance measures (Alam & Ray-Bennett, 2021; Mizutori, 2020; Ray-Bennett & Shiroshita, 2019) from the impact of a disaster. Each death is counted only once (Cook, 2019; Price, 2021). However, it is also argued that some deaths are unavoidable. Unavoidable deaths occur due to natural causes and poverty-related diseases or other causes and are often termed excess deaths. I return to these discussions in detail in Chap. 2, followed by evidence in the subsequent chapters. It is posited that national and local disaster managers and policymakers of lower-middle and low-income countries should focus on avoidable disaster deaths so that finite resources can be spent judiciously to save lives during disasters.

The avoidable disaster deaths approach is aimed at being a sense-making operational tool. It amplifies the importance of putting people and their health first during disasters. It also focuses on social justice and the equitable distribution of finite resources for high-risk and vulnerable households in the planning and operation of disaster response programmes or interventions. This approach is rooted in reducing *immediate* direct and indirect disaster deaths. As such, policy planners and decision-makers cannot overlook human mortality and morbidity because both are detrimental to growth and sustainability.

In light of this context, the aims of this book are three-fold:

(i) to help policymakers, practitioners and citizens make better decisions to save lives and reduce injuries from disasters by raising awareness of both avoidable and unavoidable disaster deaths;
(ii) to raise awareness of the circumstances and causes that led to avoidable disaster deaths during the first wave of COVID-19 and economic lockdown; and
(iii) to learn lessons from these circumstances to improve disaster responses to reduce avoidable disaster deaths in the next disaster be it biological, natural, or technological.

Defining Disaster: Terms of Reference

In 2015 the Sendai Framework (successor to the Hyogo Framework for Action 2005–2015) set four priorities for action and seven global targets. Of these seven Global Targets, the first two are relevant in the context of this study, and they are:

> Global Target A: "Substantially reduce global disaster mortality by 2030" and
> Global Target B: "Substantially reduce the number of affected people globally by 2030" (UN, 2015, p. 9).

Reducing disaster deaths is a high priority for the UN Member States. More than 180, including India, have ratified the Sendai Framework, and are currently working towards meeting these two targets by 2030. Although the Sendai Framework does not define disaster, it acknowledges for the first time that it can be caused by a 'biological hazard' in addition to hydrometeorological and geological hazards or induced by human processes (environmental degradation and technological hazards) (UN, 2015, p. 3/25).

In 2020 (five years after the launching of the Sendai Framework), the UNDRR and International Science Committee's (ISC) Technical Report on *Hazard, Definition and Classification Review* expanded the hazard list to include 302 hazards grouped into eight clusters: meteorological and hydrological hazards, extraterrestrial, geohazards, environmental hazards, chemical hazards, **biological hazards** [my emphasis], technological hazards and societal hazards (UNDRR and ISC, 2020). The Technical Report provides the definition of 'disaster' adopted by the UN General Assembly in Annex 3 which is:

> **A serious disruption of the functioning of a community or a society** [my emphasis] at any scale due to hazardous events interacting with conditions of exposure, vulnerability and capacity, leading to one or more of the following: human, material, economic and environmental losses and impacts. (UNDRR and ISC, 2020, p. 54)

This definition is a slight variation of the previously provided by UNDRR in 2009. Disaster is:

> **A serious disruption of the functioning of a community or a society** [my emphasis] involving widespread human, material, economic or environmental losses and impacts, which

Table 1.1 Typology of organisations and actors

Organisations	Actors
Primary	Category 1 responders and affected community
Secondary	Category 2 responders
Tertiary	Category 3 responders

Source Ray-Bennett (2018)

exceeds the ability of the affected community or society to cope using its own resources. (UNISDR, 2009a)[2]

The latest definition adopted by the UN General Assembly introduces the 'scale' of the hazard (local, regional and global) and interactions with conditions of three quantitative measures, including exposure, vulnerability and capacity. However, it drops the sentence on the "ability of the affected community or society to cope using its own resources" from the 2009 version. It is difficult to know why this is the case. Although both definitions are not prescriptive, this study draws attention to the highlighted text "**A serious disruption of the functioning of a community or a society**." For this study, I understand 'community or a society' as primary, secondary and tertiary actors and organisations who respond directly and indirectly during and after a disaster (see Table 1.1).

Primary organisations are the first responders. They are also known as Category 1 responders by the Civil Contingency Act in the UK (Ray-Bennett, 2018). They have specified responsibilities—"risk management, emergency planning, business continuity of the responder itself, and warning and informing the public" (Walker & Broderick, 2006, p. 81). Primary organisations also include the affected communities or 'at risk' communities (Ray-Bennett, 2018; Wisner et al., 2004). The primary responders in the context of this study are emergency services, including fire and rescue, police, ambulance, grassroots health workers or paramedics, doctors, nurses, and affected individuals and households. They provide the first response to save lives or avoid disaster deaths at home, in communities, and at hospital and non-hospital set-ups.

Secondary Organisations are category 2 responders. They are decision-making organisations at national and state levels. They undertake many roles and responsibilities, which can be categorised as auxiliary, alleviating and collateral (Ray-Bennett, 2017, 2018). Auxiliary organisations interact with the primary organisation before a disaster climate (Ray-Bennett, 2017, 2018; Toft and Reynolds, 2005). They make decisions, raise disaster funds, coordinate relief activities, and share information (among other things) with primary organisations to enhance response. Examples include the National Disaster Mitigation Authority in India, which works closely with the state-level disaster management authorities (Srivastava, 2009; Ray-Bennett, 2017, 2018). Alleviating organisations (Toft & Reynolds, 2005) complement the primary and auxiliary responders, businesses, and the public by generating valuable

[2] UNISDR (United Nations International Strategy for Disaster Risk Reduction) has been renamed to UNDRR in 2020.

early warnings for decision-making and planning for response (Ray-Bennett, 2018). Examples include the Center for Disease Dynamics, Economics and Policy and the Press Information Bureau. Collateral organisations (Lalonde, 2011, p. 450) are "new form of [secondary] organisations that do not replace the usual operational structure but co-exist and contribute to the problems that are non-routine". For example, under the auspices of the Prime Minister in New Delhi, India, the Disaster Mitigation Authority exists solely to address disaster management and mitigation issues. Other examples include the Environment Agency in the United Kingdom and the Federal Emergency Management Authority in the United States (Ray-Bennett, 2018).

Tertiary organisations are category 3 responders. These organisations are multilateral or global organisations formed between "three or more nations to work on issues that relate to all of the countries in the organization" (Global Energy Network Institute, (GENI), 2014). Examples include the UNDRR—a dedicated secretariat established by the UN General Assembly in December 1999 for disaster risk reduction. Other examples include the World Health Organisation (WHO), and the World Bank—amongst others.

In light of this discussion, 'disaster' is understood as a "serious disruption of the functionality" of the primary responders, especially the emergency services for the socio-economically vulnerable sections of the population due to the shock from the virus, inadequate disaster risk governance and economic lockdown implemented by the secondary organisations/responders. Furthermore, the serious disruptions of the services of primary responders and the community's inability to cope with the economic lockdown intersected in a complex way, leading to avoidable disaster deaths and scarcity at local and national levels in India. This is demonstrated in Chaps. 5 and 6.

The UNDRR's definitions of 'disaster' (2009, 2020) help understand and map the "serious disruption of the functioning" of the primary and secondary responders. However, it is not specific about how many deaths are required to call an event a disaster. The latest definition includes the 'scale' of the hazard and three measurable indicators, but it lacks clarity in this regard. In this light, the Centre on the Epidemiology of Disasters (CRED), which hosts the International Disaster Database (EM-DAT[3]), the definition of disaster is helpful. CRED defines an event as a disaster if at least one of the following criteria is fulfilled: ten or more people are reported killed; 100 or more people are reported affected, injured, and/or homeless; there has been a declaration of a state of emergency; or there has been a call for international assistance (Paul, 2021; Tzvetkova, 2017).

Considering the criteria identified by UNDRR, UN General Assembly and CRED to call an event a 'disaster', it can be argued that COVID-19 was a **disaster** on a global scale when this study began on 23 March 2020. By 24 April 2020, COVID-19 had crossed 210 countries, affected 2,734,102 people, and killed 191,189 (Huang et al., 2020; Worldometer, 2020a, 2020b, 2020c), with a mortality rate of 3.4%, as reported by WHO on 5 March 2020, and still ongoing. Considering this, the working

[3] https://www.emdat.be/.

definition of a 'disaster' from the perspective of avoidable disaster deaths is proposed as:

> A serious disruption of the functioning of primary and secondary disaster responders and services, destruction and disruption of health facilities and services at national and local levels from the impact of a hazard (e.g. biological) and inadequate disaster risk governance, leading to deaths of ten or more people, 100 or more people are reported affected, injured, and/or homeless and exceeds the ability of the affected community or society to cope using its own resources.
>
> (adapted from UNDRR, 2009b; UN, 2015; Tzvetkova, 2017, and based on the evidence in the following chapters).

The Virus

On 11 March 2020, the WHO declared COVID-19 a pandemic. A disease is considered a pandemic when it occurs worldwide (WHO, 2020b), or over a very wide area, crossing international boundaries and usually affecting many people (Doshi, 2011; WHO, 2020c, 2020d). COVID-19 is an infectious disease caused by a newly discovered coronavirus, identified by the WHO as SARSCoV-2 (Li et al., 2020). The virus first emerged as a series of pneumonia cases of unknown cause in December 2019 in Wuhan, China, with clinical presentations greatly resembling viral pneumonia. Deep sequencing analysis from lower respiratory tract samples indicated a novel coronavirus, which was named 2019 novel coronavirus (2019-nCoV/COVID-19) (Huang et al., 2020). The SARSCoV-2 may have jumped from wild animals to humans, but its exact origin has not been established with various suggestions. Sridhar (2022, pp. 30–38), in her book *Preventable: How Pandemic Changed the World & How to Stop the Next One*, identified several possible origins, including the 'cold chain hypothesis' [this theory postulates that bits of the virus could have been frozen into food in Place A (e.g., Italy) and then sent to Place B (Wuhan)]; the virus was bioengineered as a weapon and released intentionally; the virus infected an intermediary mammal (such as pigs, camels, snakes civets, pangolins), spread, and evolved within that species, and then jumped into humans when they had close contact; the bats infected a human with a precursor to SARS-CoV-2, and then the virus developed through a natural selection process among humans—among others (for more detail, please see Sridhar, 2022, pp. 30–38).

Before the severe acute respiratory syndrome (SARS) epidemic, only two human coronaviruses had been characterized (HuCoV-229E and HuCoV-OC43). Both usually cause a mild upper respiratory tract infection (Li et al., 2020). The current strain differs from previous SARS (SARS-CoV: November 2002 to July 2003) and Middle East Respiratory Syndrome coronavirus (MERS-CoV: 2012). Although most human coronavirus infections are mild, the epidemics of the two betacoronaviruses, SARS-CoV, and MERS-CoV have caused more than 10,000 cumulative cases in the past two decades, with mortality rates of 10% for SARS-CoV and 37% for MERS-CoV; (Huang et al., 2020) but nothing compared to COVID-19 as mentioned earlier because of its reach and mortality rate (Huang et al., 2020; Sridhar, 2022; Worldometer, 2020a, 2020b, 2020c).

The COVID-19 virus affects different people in different ways. It is highly infectious and pathogenic, causing human-to-human transmission. While writing this book in 2021, COVID-19 was considered a respiratory disease, and most infected people will develop mild to moderate symptoms and recover without requiring special treatment. Some people become infected but do not develop symptoms or feel unwell. Most people (about 80%) recover from the disease without special treatment. Around one out of every six people who get COVID-19 becomes seriously ill and develops difficulty breathing (WHO, 2020a; MOHFW, 2020). People who have underlying medical conditions and those over 60 years old have a higher risk of developing severe disease and death. The main routes of transmission are respiratory droplet transmission and contact transmission. Close contact with patients with SARSCoV-2 infection and asymptomatic infection is the primary way to cause disease. These droplets land on objects and surfaces around the person. Other people then catch COVID-19 by touching these objects or surfaces and then touching their eyes, noses, or mouths. Studies suggest that coronavirus may persist on surfaces for several hours or days. This may vary under different conditions (e.g., the type of surface, sunlight, temperature, or humidity of the environment). It has also been reported that coronavirus can be transmitted through aerosols, the faecal-oral route, and suspected mother-to-infant transmission (Li et al., 2020; MOHFW, 2020).

Whilst writing this book in 2020, the incubation period of SARSCoV-2 infection was 1 to 14 days, generally 3 to 7 days ((Li et al., 2020). The clinical manifestations are asymptomatic or include fever, fatigue, and dry cough; a few patients have upper respiratory tract symptoms, such as nasal obstruction, runny nose, and sore throat; and a few patients have gastrointestinal symptoms, such as abdominal discomfort, nausea, vomiting, stomach-ache, and diarrhoea (Li et al., 2020). According to Rothan and Byrareddy (2020), the symptoms of COVID-19 infection appear after an incubation period of approximately 5.2 days. The period from the onset of COVID-19 symptoms to death ranged from 6 to 41 days, with a median of 14 days. This period depends on the patient's age and the patient's immune system status. It was shorter among patients > 70 years old compared with those under 70. The most common symptoms at the onset of COVID-19 illness are fever, cough, and fatigue, while other symptoms include sputum production, headache, haemoptysis, diarrhoea, dyspnoea, and lymphopenia.

Curbing the Virus

In 2020, the WHO rolled out several interventions to help curb the spread of the virus, such as hand hygiene, respiratory etiquette, social distancing, crowd control, and lockdown. The overarching goal was for all countries to control the pandemic by slowing the transmission and reducing mortality associated with COVID-19 (WHO, 2020e). The global strategic objectives were as follows:

(i) Mobilise all sectors and communities to ensure that every sector of government and society takes ownership of and participates in the response and prevention of cases through hand hygiene, respiratory etiquette, and individual-level physical distancing.
(ii) Control sporadic cases and clusters and prevent community transmission by rapidly finding and isolating all cases, providing them with appropriate care, tracing, quarantining, and supporting all contacts.
(iii) Suppress community transmission through context-appropriate infection prevention and control measures, population-level physical distancing measures, and appropriate and proportionate restrictions on non-essential domestic and international travel.
(iv) Reduce mortality by providing appropriate clinical care for those affected by COVID-19, ensuring the continuity of essential health and social services, and protecting frontline workers and vulnerable populations.
(v) Develop safe and effective vaccines and therapeutics that can be delivered at scale and accessible based on need (WHO, 2020e).

National governments have implemented several strategic objectives, including travel bans, social distancing, hand washing, mass testing, contact tracing, national emergencies, and economic lockdowns (Sridhar, 2022). Although these interventions have proven effective in curbing the spread of the virus in China, Singapore, and the UK in 2020, national lockdowns (Sridhar, 2022)—including the closure of all economic activities—have had a debilitating effect on the poorest and vulnerable globally and in low- and high-middle income countries, in particular (see Chap. 6).

In the first phase, Oxfam (quoted in BBC, 2020) warned that the economic fallout from COVID-19 could potentially be more severe than the health crises, posing real threats to the UN's Sustainable Development Goals. Furthermore, it forecasted that by the time the pandemic is over, half of the world's population of 7.8 billion people could live in poverty. About 40% of the new poor could be concentrated in East Asia and the Pacific, with about one-third in both Sub-Saharan Africa and South Asia (BBC, 2020).

India and Lockdown

In South Asia, India was predicted to experience the impact of this biological disaster disproportionately (Klein et al., 2020). According to these authors, around 300–400 million Indians were likely to be infected by July 2020. There was a likelihood of recurrence in November/December, and the national lockdown will "increase hunger and reduce the population resilience for handling the infection peak" (Klein et al., 2020).

India is home to 1.37 billion people (Worldometer, 2020a, 2020b, 2020c). Although the extreme poverty level dropped from 46% to an estimated 13.4% over the two decades before 2015, India is still home to 176 million poor people (The

World Bank Annual Report, 2019). To curb the spread of the virus in the second-largest populous country in the world, on 24 March 2020, the National Disaster Management Authority (NDMA) invoked the 'Disaster Management Act of 2005'. To my knowledge, India was the only country that enforced this Act in 2020. As such, COVID-19 is referred to hereafter as a disaster or biological disaster in the context of this study. Given my background in DRR and development, India became a unique case to study avoidable disaster deaths from the impact of this biological disaster and economic lockdown.

As per Section 6(2)(i) of the 'Disaster Management Act, 2005', the National Disaster Management Authority directed all Ministries/Departments of the Government of India, State Governments as well as State authorities to adopt measures to enforce social distancing in India (NDMA, 2020). On 24 March, the Ministry of Home Affairs also released an order calling for a lockdown across the country for 21 days, emphasising the "need for consistency in the application and implementation of various measures across the country" (Ministry of Home Affairs, 2020a).

On 14 April 2020, the National Disaster Management Authority directed the Central and State Governments to extend the 'strict social distancing measures' till 3 May 2020. At the end of the initial 21 days of the lockdown period, the Prime Minister of India, Mr Narendra Modi, announced the extension of the lockdown (Press Information Bureau (PIB), 2020a). According to the Guidelines of the Government of India released on 24 March 2020, it is stated that except for a few offices, commercial, private and industrial establishments, educational, training and research institutions and places of worship were to be closed for the lockdown period. Hospitality and all transport services were suspended, and all social, political, sports, academic and other institutions were barred. A maximum of 20 persons were permitted to attend a funeral, and all persons who arrived in India after 15 February 2020 and those directed by healthcare personnel to do so were required to remain under quarantine. If anybody fails to observe this, they are liable for legal action under Section 188 of the Indian Penal Code. However, hospitals, veterinary hospitals, pharmacies, chemists, ambulances, nursing homes and all related medical establishments were to remain functional. Transportation for all medical personnel, para-medical staff and other hospital support services was permitted (Ministry of Home Affairs, 2020b).

The Ministry of Home Affairs also invoked provisions of the Essential Commodities Act 1955 to ensure the availability of essential goods. The Act provides measures such as "fixing of stock limits, capping of prices, enhancing production, inspection of accounts of dealers and other such actions" (PIB, 2020b, p. 1). The lockdown policy continued to evolve over the months, and this is discussed further in Chap. 5.

India Today (2020) described the lockdown in the early months as the 'strictest' in the world. The concerted effort to maintain social distancing (a preventative measure) in the most populous country on the planet was unparalleled in any other country. Incidents of police brutality were reported across the country, and citizens were not spared even if they were on the streets to purchase essential commodities (Asthana, 2020). An extreme example of this was found in late March in West Bengal, where a 32-year-old man succumbed to injuries sustained from police brutality as he left home to purchase milk for his family (The Wire, 2020). The Commonwealth Human Rights

Initiative (2020) petitioned the National Human Rights Commission to investigate the deaths by police brutality of 15 individuals between the period of 25 March and 30 April 2020.

The 'strictest' lockdown also triggered a "humanitarian disaster of unprecedented proportions" (Sengupta and Jha, 2020, p. 152) for socio-economically vulnerable groups, including migrant workers. The first lockdown was announced in haste on 24 March 2020. After a curfew of 24 h, the Ministry of Home Affairs announced the lockdown for 21 days in less than 7 h. This gave little time for those who were far away to return home. Migrant workers are a case in point.

Circa 90% of the Indian labour force is employed in the unorganised sector, and millions work far from their rural homes in urban areas. These workforces faced tremendous pressure from landlords to pay rent or face eviction (Joshi, 2020) despite orders from the Government of India forbidding landlords to evict labourers and students (Ministry of Home Affairs, 2020b). The economic lockdown led to the loss of livelihoods and the inability to pay house rent, which in turn led to widespread hunger and destitution amongst the migrant workers. As a result, millions of migrant workers decided to return home. They walked hundreds of kilometres on foot, carrying their belongings, children and sometimes elderly on their backs. These images caught global attention.

Although the migrant crisis is a bone of contention, the Ministry of Labour and Employment during the parliament (Lok Sabha) question time, however, acknowledged: "Covid-19 has resulted in a large number of migrant workers going back to their native places and loss of jobs has also been observed in India during this period" (Government of India, 2020). The causes and circumstances surrounding avoidable disaster deaths of the migrant workers and others will be resumed in depth in Chap. 6 through the components of avoidable preventable, amenable, and governance-related deaths. The purpose of presenting these cases is not to criticise the disaster management and emergency health systems but instead to showcase the bigger picture and learn lessons from this pandemic disaster to improve disaster risk governance for the disaster and emergency health systems in lower-middle—and low-income settings. The components of avoidable disaster deaths provide a means of doing so.

Data Collection in India

It was whilst watching and reading about the horrors of economic lockdown and its effect on migrant workers and others that this prospective project was conceived. To answer, 'How many avoidable disaster deaths were occurring from the impact of COVID-19 and economic lockdown, and what were the circumstances and causes of these deaths' four indicators were identified: (i) preventable deaths, (ii) amenable deaths, (iii) disaster risk governance-related deaths; and (iv) unavoidable deaths.

These indicators are discussed at length in Chap. 3, followed by evidence in Chaps. 4–6. To collect data for this study, ethics approval was obtained from the University of Leicester School of Business Research Ethics Committee.[4]

The project coordinator and I led a team of 12 reviewers who reviewed the content of nine reputable English e-newspapers, three digital news services, and one e-magazine every day for nine months (11 March—31 December 2020). The only exceptions were The Sentinel and The Tribunal, whose review began in July 2020. It was observed that the selective mainstream national media did not cover news from the northeastern regions of India very well. To represent these States, the Sentinel and The Tribunal were included for review.

The frequency of data entry for avoidable disaster deaths was high in the early months of COVID-19. However, as the months progressed, the reporting of fatalities within these sources continued to fall. By December, the content review team were unable to find any stories. As such, the content review was stopped. This means that this study did not cover the second phase of COVID-19 in early 2021, which again shocked the global community through the sheer number of deaths and lack of oxygen supply—among other things.

To understand unavoidable deaths[5] six relevant reports published by the Government of India's different departments were reviewed (see Table 1.2). This included four reports published by the Office of the Registrar General and Census Commissioner of India and two by the National Crime Records Bureau. Data related to disaster risk governance (DRG) began on 30 June and was concluded on 31 December. For this, the websites of the Press Information Bureau, the Ministry of Health and Family Welfare, The Ministry of Home Affairs and the National Disaster Management Authority were reviewed.

Conclusion: Novelty and Limitations

India is the seventh largest country in the world, extending from the Great Himalayas in the north to the Bay of Bengal in the east and the Arabian Sea in the west. India consists of 28 States and eight Union Territories (Know India, 2024). Given India's vast geography and cultural diversity, the data reviewed and collected for this study are grossly underrepresented. Furthermore, the exclusion of hundreds of vernacular newspapers published daily at national and local levels in 28 States and eight Union Territories was beyond the remit of this study; as such, this is a significant limitation of this study.

India is diverse culturally and linguistically. Also, the development indicators are not uniform. Since most content reviewers were international, only English newspapers were chosen for the review. Despite these limitations, this study has a few merits.

---

[4] Ethics reference: 27314-nsrb1-ss/bu:innovation,technology&ops (24.07.2020).

[5] These as well as older editions are available on the websites of the concerned departments.

Table 1.2 Sources for data

E-newspapers	Digital news services	E-magazine	Reports	Websites
The Times of India; The Guardian; The New York Times; The New Indian Express; The New Indian Express Odisha; The Hindu; The Tribune; The Sentinel; and Sambad	Bloomberg; BBC News; The Wire; and Scroll	India Today	Published by the Office of the Registrar General and Census Commissioner of India – Cause of Deaths in India 2010–2013 – Cause of Deaths in India Statistics 2007–2013 – Sample Registration System (SRS) Statistical Report 2017 – Report on the Medical Certification of Cause of Death 2017 Published by the National Crime Records Bureau – Accidental Deaths and Suicides in India 2018	Press Information Bureau; The Ministry of Health and Family Welfare; The Ministry of Home Affairs; The National Disaster Management Authority

It provides a snippet of avoidable disaster deaths and their circumstances in 2020. Also, it gives a novel avoidable disaster deaths approach to make sense of these deaths, and in doing so, this study is a significant departure from the other studies, including databases that emerged in 2020. When the study began in 2020, five databases were found to provide updates on the number of COVID-19 deaths, infected cases, and recovered cases from COVID-19. A brief overview of these databases is provided here:

Worldometers is a free reference website that provides up-to-date information for each country. Official reports, government informational channels, and reliable local media sources, including live streams, collate and verify over 5,000 sources of statistical data, which are then validated by a team of analysts to ensure accuracy before publishing (Worldometers, 2020a, 2020b, 2020c).

John Hopkins University initiated an online tracking platform similar to Worldometers. Reputable sources, including official government websites and trusted media, collect and verify data, and the platform is updated daily (John Hopkins University and Medicine, 2020[6]).

The Imperial College London has a database for 29 countries covering the number of cases and deaths recorded. In partnership with YouGov, Imperial College London also studies the changes taking place in their selective countries (Imperial College London, 2020). Government interventions via policy changes, such as country curfews and lockdowns, are recorded per country. Citizens' behaviour in abiding by these changes and the recommended behaviours related to social distancing are analysed and published on the databases.

The Government of India's Ministry of Health and Family Welfare has a similar general database interface to Worldometers. It displays the number of cases, direct deaths, recovered cases, and persons migrating. It also provides an interactive map of India with accompanying graphs that break down the data by State (Ministry of Health and Family Welfare, 2020). The map is colour-coded to show the worst-affected Indian states.

A small group of volunteers who were general medicine physicians were found collecting data on direct deaths in the State of Kerala. This group "checked the district editions of seven local newspapers and followed five news channels" daily over nine months (28 March–23 December 2020) (Biswas, 2020).

These databases are unique and insightful and generate data on the number of direct deaths, infected cases and recovered cases from COVID-19. On the other hand, this study is about avoidable disaster deaths from the impact of COVID-19 and the economic lockdown. Therefore, this study is novel. Like this study, two databases[7] were found in 2020:

The Save Life Foundation maintained an online tracking platform named 'Road Crash Deaths During Lockdown' in India. Its sources included over 30 print and media sources that are accessible online, including national, regional, and vernacular media reporting in English, Hindi, and other vernacular languages (Save Life, 2020a, 2020b). Fatalities are recorded into three categories: migrant deaths, general road user deaths, and essential road user deaths. Crashes and deaths are segregated by State, causes of death, and mode of transport are recorded in various tables.

Thejesh (2020***)*** is another public database that captured data related to 'Non-Virus Deaths' due to the lockdown. The sources include newspapers, magazines and YouTube videos published by news agencies. The deaths in this database are categorised as: "starvation and financial distress, exhaustion, accidents during migration, lack or denial of medical care, suicides, police brutality, crimes and alcohol-withdrawal" (Thejesh, 2020). Cases of deaths in which conflicting reports emerged or that required more information are put into the 'unclassified' category. The database contains a table with information regarding the number of deaths,

[6] The tracking was discontinued on 3 October 2023 (see https://coronavirus.jhu.edu/data).

[7] Of these deaths, we may find that fatalities fall within the category of avoidable deaths in general and preventable, amenable and disaster risk governance deaths in specific.

reasons for death, location of death, source of information, and the name and occupation of the individual(s) who died.

Although these two databases are insightful and relevant, this study is distinctive. First, this book focuses on capturing 'avoidable disaster deaths' and not just 'disaster deaths' from the impact of COVID-19 and the economic lockdown under the ambit of the Sendai Framework's Global Target A (reducing disaster deaths—direct/indirect and immediate/delayed) and public health/epidemiology (circumstances and causes). To do so, I redefine the mainstream 'disaster' definition and introduce the novel approach 'avoidable disaster deaths' to make sense of which deaths are avoidable and unavoidable in a disaster climate (see the section Defining Disaster). My approach is innovative and multidisciplinary. It spans DRR, public health and epidemiology, human geography, and crisis and risk management, representing a departure from the two databases (Save Life and GN) mentioned above. In so doing, I anticipate that my novel approach will open new lines of inquiry in disaster risk reduction and disaster risk management policy on how to achieve the Sendai Framework's Goals for 'reducing disaster deaths' and the number of 'people affected by disasters' by 2030 in lower-middle and low-income countries. As such, this book is novel.

This book also brings 15 descriptive pieces of evidence related to disaster risk governance and 29 descriptive cases of avoidable disaster deaths to the forefront. The descriptive cases capture the circumstances that led to avoidable disaster deaths so that lessons can be learned to improve disaster and emergency health system responses in India and beyond. This is missing in the abovementioned studies. As mentioned at the beginning of this chapter, the burden of these deaths was borne mainly by the socio-economically vulnerable sections of society. The deceased are not here to tell their stories, but we are. We have lost far too many people in this historic, perhaps once-in-my-lifetime global disaster. Lessons ought to be learnt, and the first step towards this is the transformative potential of remembering them, as succinctly put by the wise Roman poet Virgil (19BC): "*No day shall erase you from the memory of time.*"

I hope that the avoidable disaster deaths approach introduced in this book serves as a remembrance tool to usher in positive change. To bring about positive change, it is essential to know why people die in disasters, who those people are, what causes and circumstances lead to their deaths, and how we can reduce them. I explore these questions in the following two chapters.

References

Alam, E., & Ray-Bennett, N. S. (2021). Disaster risk governance of local level administration in Rangamati, Bangladesh. *International Journal for Disaster Risk Reduction, 59*(2021), 1–13. https://doi.org/10.1016/j.ijdrr.2021.102220

Asthana, N. C. (2020). Amidst a lockdown, why must cops wield the Lathi with such impunity and callousness? *The Wire,* 27 March. Available at: https://thewire.in/government/lathi-police-brutality-coronavirus. Accessed on September 7, 2020.

BBC. (2020). *Virus could push half a billion people into poverty.* BBC News. Available at: https://www.bbc.com/news/business-52211206. Accessed on January 11, 2022.

Biswas, S. (2020). *India coronavirus: How a group of volunteers 'exposed' hidden Covid-19 deaths.* BBC News. Available at: https://www.bbc.com/news/world-asia-india-54985981. Accessed on January 10, 2021.

Commonwealth Human Rights Initiative. (2020). *CHRI petitions NHRC to investigate 15 deaths following alleged police excesses during COVID-19 lockdown.* Available at: https://www.humanrightsinitiative.org/publication/chri-petitions-nhrc-to-investigate-15-deaths-following-alleged-police-excesses-during-covid19-lockdown. Accessed on August 7, 2020.

Cook, A. (2019). *Avoidable mortality in the UK: 2017.* Office of National Statistics (ONS). https://www.ons.gov.uk/peoplepopulationandcommunity/healthandsocialcare/causesofdeath/bulletins/avoidablemortalityinenglandandwales/2017. Accessed on April 19, 2022.

Dattani, S., Spooner, F., Ritchie, H., & Roser, M. (2023). Causes of death. *OurWorldinData.org.* Available at: https://ourworldindata.org/causes-of-death. Accessed on October 27, 2024.

Doshi, P. (2011). The elusive definition of pandemic influenza. *Bulletin of the World Health Organization, 89*(7), 532–538. https://doi.org/10.2471/blt.11.086173

GENI (Global Energy Network Institute). (2014). *Multilateral organizations.* Global Energy Network Institute. http://www.geni.org/globalenergy/library/organizations/index.shtml#global. Accessed on January 29, 2015.

Government of India. (2020). *Lok Sabha unstarred question no 100, answered on 14.09.2020: Job loss and reverse migration.* Available at: http://164.100.47.194/Loksabha/Questions/Qresult15.aspx?qref=16381&lsno=17. Accessed on September 16, 2020.

Huang, C., Wang, Y., Li, X., Ren, L., Zhao, J., Hu, Y., Zhang, L., Fan, G., Xu, J., Gu, X., Cheng, Z., Yu, T., Xia, J., Wei, Y., Wu, W., Xie, X., Yin, W., Li, H., Liu, M., & Xiao, Y. (2020). Clinical features of patients infected with 2019 novel coronavirus in Wuhan, China. *The Lancet, 395*(10223), 497–506. https://doi.org/10.1016/s0140-6736(20)30183-5

Imperial College London. (2020). *COVID-19 behaviour tracker, Institute of Global Health Innovation.* Available at: https://public.tableau.com/app/profile/ighi/viz/ICLYouGovCovid-19Tracker_V0_3/1Specificbehaviourbycountry. Accessed on April 20, 2020.

India Today Web Desk. (2020). India implements strictest lockdown in the world, lags in testing: Expert. *India Today*, April 10. Available at: https://www.indiatoday.in/india/story/india-implements-strictest-lockdown-in-the-world-lags-in-testing-expert-1665604-2020-04-10. Accessed on September 5, 2020.

John Hopkins University and Medicine. (2020). *Tracking.* Coronavirus Resource Centre. Available at: https://coronavirus.jhu.edu/data. Accessed on May 31, 2020.

Joshi, A. (2020). No relief: Pay rent or leave, landlords tell migrant workers. *The Tribune,* 19 May. Available at: https://www.tribuneindia.com/news/jalandhar/no-relief-pay-rent-or-leave-landlords-tell-migrant-workers-86539. Accessed on September 8, 2020.

Kayyem, J. (2022). When disaster strikes, what you don't know might kill you. *The Harvard Gazette,* 8 April. Available at: https://news.harvard.edu/gazette/story/2022/04/most-died-in-blizzard-of-78-from-co2-poisoning/. Accessed on April 20, 2022.

Klein, E., Lin, G., Tseng, K., Schueller, E., Kapoor, G., & Laxminarayan, R. (2020). *COVID-19 for India updates.* The Center for Disease Dynamics, Economics and Policy. Available at: https://onehealthtrust.org/wp-content/uploads/2020/03/covid19.indiasim.March23-2-eK.pdf. Accessed on June 15, 2020.

Know India. (2024). *States and union territories.* India.Gov.In. Available at: https://knowindia.india.gov.in/states-uts/. Accessed on October 21, 2024.

Lalonde, C. (2011). Managing crisis through organisational development: A conceptual framework. *Disasters, 35*(2), 443–464. https://doi.org/10.1111/j.1467-7717.2010.01223.x

Li, Y., Guo, F., Cao, Y., Li, L., & Guo, Y. (2020). Insight into COVID-2019 for Pediatricians. *Pediatric Pulmonology, 55*(5), E1–E4. https://doi.org/10.1002/ppul.24734

Mallapur, C. (2019). While Sri Lanka grows richer, India remains a lower-middle income nation. Scroll.in, 5 July. Available at: https://scroll.in/article/929406/while-sri-lanka-has-grown-richer-india-remains-a-lower-middle-income-nation#:~:text=India%20continues%20to%20be%20a%20lower-middle-income%20country%20along,countries%20by%20income%20levels%2C%20released%20on%20July%201. Accessed on October 13, 2024.

Ministry of Health and Family Welfare (MOHFW). (2020). *MOHFW: Home*. Available at: https://www.mohfw.gov.in/. Accessed on April 20, 2020.

Ministry of Home Affairs. (2020a). *Order (No. 40-3/2020-DM-I(A))*. Available at: https://www.mha.gov.in/sites/default/files/MHAorder%20copy_0.pdf. Accessed on September 5, 2020.

Ministry of Home Affairs. (2020b). *Annexure to Ministry of Home Affairs Order No. 40-3/2020-D dated 24.03.2020*. Available at: https://www.mha.gov.in/sites/default/files/MHAorder%20copy_0.pdf. Accessed on September 5, 2020.

Mizutori, M. (2020). Reflections on the Sendai framework for disaster risk reduction: Five years since its adoption. *International Journal of Disaster Risk Science, 11*, 147–151.

NDMA (National Disaster Management Authority). (2020). *Order (No. 1-29/2020-PP (Pt. II))*. Available at: https://ndma.gov.in/images/covid/ndmaorder240320.pdf. Accessed on September 05, 2020.

Paul, B. K. (2021). *Disaster deaths: Trends, causes and determinants*. Routledge.

PIB (Press Information Bureau). (2020a). *Text of PM's address to the Nation*. Government of India. Available at: https://pib.gov.in/PressReleseDetail.aspx?PRID=1614215. Accessed on September 05, 2020.

PIB (Press Information Bureau). (2020b). *MHA directs all States/Uts to ensure strict compliance of lockdown measures to fight COVID—19 and not allow any social/religious gathering/procession*. Available at: https://pib.gov.in/PressReleseDetail.aspx?PRID=1612928. Accessed on September 7, 2020.

Price, M. (2021). Avoidable mortality in the UK: 2019. *Statistical Bulletin*. Available online: www.ons.gov.uk

Ray-Bennett, N. S. (2006). *Caste, class and gender: Multiple disasters and women-headed households in an Oriya Village*. Ph.D. Thesis, Department of Sociology, University of Warwick, UK.

Ray-Bennett, N. S. (2009). *Caste, class and gender in multiple disasters: The experiences of women-headed household's in an Oriya Village*. VDM Verlag.

Ray-Bennett, N. S. (2017). Disasters, deaths, and the Sendai goal one: Lessons from Odisha, India. *World Development, 103*(2018), 27–39. https://doi.org/10.1016/j.worlddev.2017.10.003

Ray-Bennett, N. S. (2018). *Avoidable deaths: A systems failure approach to disaster risk management*. Springer Nature.

Ray-Bennett, N. S., & Shiroshita, H. (2019). Disasters, deaths and the Sendai framework's target one: A case for systems failure in Hiroshima landslide 2014, Japan. *Disaster Prevention and Management, 28*(6), 764–785. https://doi.org/10.1108/DPM-09-2019-0302

Rothan, H. A., & Byrareddy, S. N. (2020). The epidemiology and pathogenesis of coronavirus disease (COVID-19) outbreak. *Journal of Autoimmunity, 109*(102433), 102433. https://doi.org/10.1016/j.jaut.2020.102433

SaveLIFE Foundation. (2020a). *Road crash deaths during COVID lockdown tracker*. SaveLIFE Foundation. Available at: https://savelifefoundation.org/wp-content/uploads/2020/06/ROAD-CRASH-TRACKER-KEY-DATA-POINTS-SLF.pdf. Accessed on July 16, 2020.

SaveLIFE Foundation. (2020b). *Road crash deaths during lockdown*. Savelifefoundation.org. Available at: https://savelifefoundation.org/covidtracker/. Accessed on July 16, 2020.

Sengupta, S., & Jha, M. K. (2020). Social policy, COVID-19 and impoverished migrants: Challenges and prospects in locked down India. *The International Journal of Community and Social Development, 2*(2), 152–172. Available at: https://doi.org/10.1177/2516602620933715. Accessed on September 08, 2020.

Spiegelhalter, D., & Masters, D. (2021). *Covid by numbers: Making sense of the pandemic with data*. Pelican.
Sridhar, D. L. (2022). *Preventable: How a pandemic changed the world and how to stop the next one*. Penguin.
Srivastava, S. K. (2009). Making a technological choice for disaster management and poverty alleviation. *Disasters 33*(1), 58–81. https://doi.org/10.1111/j.1467-7717.2008.01062.x
The Wire. (2020). 22 migrant workers, kin have died trying to return home since the lockdown started. *The Wire*, 30 March. Available at: https://thewire.in/rights/coronavirus-national-lockdown-migrant-workers-dead. Accessed on March 30, 2020.
Thejesh, G. N. (2020). *Non virus deaths*. Available at: https://thejeshgn.com/projects/covid19-india/non-virus-deaths/. Accessed on July 16, 2020.
The World Bank Annual Report. (2019). *Ending poverty, investing in opportunity (English)*. World Bank Group. http://documents.worldbank.org/curated/en/156691570147766895/The-World-Bank-Annual-Report-2019-Ending-Poverty-Investing-in-Opportunity. Accessed on December 22, 2024.
Toft, B., & Reynolds, S. (2005). *Learning from disasters: A management approach* (3rd ed.). Palgrave Macmillan.
Tzvetkova, S. (2017). *Not all deaths are equal: How many deaths make a natural disaster newsworthy?* Our World in Data. Available at: https://ourworldindata.org/how-many-deaths-make-a-natural-disaster-newsworthy. Accessed on November 4, 2020.
UN (United Nations). (2015). *Sendai framework for disaster risk reduction 2015–2030*. UNDRR. Available at: http://www.unisdr.org/we/inform/publications/43291. Accessed on October 9, 2024.
UNDRR (United Nations Office for Disaster Risk Reduction) and ISC (International Science Congress). (2020). *Hazard definition and classification review: Technical report. A report developed by Sendai framework for DRR 2015–2030*. International Science Congress (ISC) and United Office for Disaster Risk Reduction, Geneva. Available online https://www.undrr.org/publication/hazard-definition-and-classification-review-technical-report. Accessed on September 20, 2023.
UNISDR (United Nations International Strategy for Disaster Reduction). (2009a). *2009 UNISDR terminology on disaster risk reduction*. UNISDR. http://www.preventionweb.net/files/7817_UNISDRTerminologyEnglish.pdf. Accessed on August 2, 2016.
UNISDR (United Nations International Strategy for Disaster Reduction). (2009b). *2009 UNISDR terminology on disaster risk reduction*. UN and UNISDR. Available at:http://www.preventionweb.net/files/7817_UNISDRTerminologyEnglish.pdf. Accessed on February 11, 2025.
Virgil. (19BC). *The Aeneid: Book IX*. MIT.edu. Available at: http://classics.mit.edu/Virgil/aeneid.9.ix.html. Accessed on July 16, 2021.
Walker, C., & Broderick, J. (2006). *The civil contingencies act 2004: Risk, resilience and the law in the United Kingdom*. Oxford University Press.
WHO (World Health Organization). (2020a). *Advice for the public: Coronavirus disease (COVID-19)*. World Health Organization. Available at: https://www.who.int/emergencies/diseases/novel-coronavirus-2019/advice-for-public. Accessed on April 21, 2020.
WHO (World Health Organization). (2020b). *Coronavirus*. World Health Organization. Available at: https://www.who.int/health-topics/coronavirus/coronavirus#tab=tab_1. Accessed on April 20, 2020.
WHO (World Health Organization). (2020c). *Coronavirus: WHO chief urges end to 'Politicisation' of virus*. World Health Organization. Available at: https://www.who.int/docs/default-source/coronaviruse/transcripts/who-audio-emergencies-coronavirus-press-conference-full-08apr2020.pdf. Accessed on April 9, 2020.
WHO (World Health Organization). (2020d). *COVID-19 strategy update*. World Health Organization. Available at: https://www.who.int/publications/m/item/covid-19-strategy-update. Accessed on April 21, 2020.

WHO (World Health Organization). (2020e). *International classification of functioning, disability and health (ICF)*. World Health Organization. Available at: https://www.who.int/classifications/international-classification-of-health-interventions. Accessed on April 21, 2020.

Wisner, B., Blaikie, P., Cannon, T., & Davis, I. (2004). *At risk: Natural hazards, people's vulnerability and disasters* (2nd ed.). Routledge.

Worldometers. (2020a). *COVID live—coronavirus statistics—Worldometer*. [online] Worldometers.info. Available at: https://www.worldometers.info/coronavirus/. Accessed on April 20, 2020.

Worldometers. (2020b). *India population (2020)—Worldometer*. [online] Worldometers.info. Available at: https://www.worldometers.info/world-population/india-population/. Accessed on April 20, 2020.

Worldometers. (2020c). *Worldometers—About Us*. Worldometers.info. Available at: https://www.worldometers.info/about/. Accessed on April 20, 2020.

Chapter 2
What Leads to Disaster Deaths?

Disaster is not for the deceased but for the surviving family members
—Survivor of the Odisha's Super-Cyclone 1999, India (Personal communication with a disaster survivor whilst I was collecting data for my PhD fieldwork between 2002 and 2003 in Odisha.)

Abstract To reduce disaster deaths, it is essential to understand: why people die in disasters? who are those people? what are the causes and circumstances surrounding those deaths? and how can disaster deaths be reduced? These questions are answered by engaging with several approaches, including disaster risk reduction, vulnerability, determinants of disaster-induced deaths, systems failure, and avoidable deaths. These approaches cross-cut several disciplines including disaster, geography, sociology, crisis and risk management, public health and epidemiology. Empirical data from secondary literature is provided wherever possible to illuminate the answers. Reducing disaster deaths and avoidable disaster deaths is a 'complex problem', and the engagement with different approaches and actors demonstrates its complexity and solutions to reduce them.

Keywords Disaster deaths · Avoidable deaths · Disaster risk reduction · Vulnerability · Determinants of disaster-induced deaths · Systems failure · Disaster risk governance

Introduction

The direct impact of disasters is damage and loss, which can be economic, non-economic, or both. I have always been fascinated by non-economic losses, especially human losses, and their impact on the livelihoods of the surviving members. I have always pondered: Why do people die in disasters? Who are those people? What are the causes and circumstances surrounding those deaths? How can we reduce them?

I have been seeking answers to these questions since the aftermath of the Super-Cyclone in 1999 in Odisha—mentioned in Chap. 1 (Ray-Bennett, 2009, 2017, 2018).

N. S. Ray-Bennett, *Avoiding Disaster Deaths*,
https://doi.org/10.1007/978-3-031-81300-9_2

This chapter presents some of the answers through the lens of four approaches: disaster risk reduction (DRR), vulnerability (Kapur, 2010; Paul, 2021), determinants of disaster-induced deaths (Paul, 2021) and systems thinking (Ray-Bennett, 2018). This chapter also introduces the concept of avoidable deaths (Price, 2021; Rutstein et al., 1976, 1983) to make a case for 'avoidable disaster deaths' in the next chapter.

Disaster Risk Reduction (DRR) Approach

Between 1998 and 2017, climate-related and geophysical disasters killed 1.3 million people, and another 4.4 billion were injured, homeless, displaced, or in need of emergency assistance (UNISDR, 2018). Although 91% of all disasters were caused by floods, storms, droughts, heatwaves, and other extreme weather events, most deaths were due to geophysical events, mainly earthquakes and tsunamis (UNISDR, 2018). According to the World Meteorological Organisation's (WMO, 2021) *Atlas of Mortality and Economic Losses,* the number of disaster deaths declined despite the fivefold increase in disasters from 1970 and 1979 to 2000 and 2009. During the same period, non-economic losses due to disasters increased mainly in high-income countries, and most disaster-related deaths occurred in lower-middle and low-income countries (WMO, 2021). This trend indicates that the global community is better at preventing deaths from disasters than in the past but the frequency and intensity of natural hazards are on the rise due to the impact of climate change. Reducing the impact of natural hazard disasters on humans and economies is currently an existential matter for sustainability, especially in lower-middle and low-income countries carrying the burden of disaster deaths (Ray-Bennett et al., 2020; UNISDR, 2015a, 2015b). So, how has this matter been addressed by the disaster risk reduction (DRR) community largely led by the UNDRR since the 1990s?

According to this approach, natural hazards[1] originate from natural systems and can cause harm and loss. One way of mitigating the effect of nature is through technology or a 'technical fix'; this thinking was dominant until 1994 (Ariyabandu & Wickramasinghe, 2003; Bryant, 2005; Ray-Bennett, 2009; Wisner et al., 2004). As a result, structural mitigation measures, such as building concrete houses, flood levies, ocean wave barriers, cyclone shelters, embankments, and dams, attained primary importance over non-structural mitigation measures, such as policies, laws, training, raising public awareness and aid—amongst others (Davis & Gupta, 1991; Ray-Bennett, 2018; Zaman, 1999). These measures acted as an approach to implicitly reduce disaster deaths. This technical approach evolved due to the mid-term evaluation of the United Nations International Decade for Natural Disaster Reduction (IDNDR) (1990–2000) in 1994 (known as the Yokohama Strategy) followed by the Hyogo Framework for Action (2005–2015) and most recently the Sendai Framework

[1] The Sendai Framework for Action defines a hazard as "A potentially damaging physical event, phenomenon […] that may cause the loss of life or injury, property damage […]" (UN, 2015. p. 3/25).

(2015–2030) (UN, 2015; UNISDR, 2005). Now there is widespread acknowledgement that hazards can include "latent conditions that may represent future threats and can have different origins: natural (geological, hydrometeorological and biological) or induced by human processes (environmental degradation and technological hazards)" (UNISDR, 2015a, p. 3/25).

Governmental organisations often use natural causes or the geophysics of a hazard to explain deaths in disasters (Ray-Bennett, 2018). This is noted by Kapur (2010) whilst reviewing the effects of 16 natural hazards[2] in India from 1977 to 2002. The geophysics of a hazard can be understood in three ways: intensity, seasonality, and spatial distribution. First, the higher the intensity of a hazard, the more likely it is to kill people. Intensity is classified as moderate or severe for 11 hazards out of the 16. Of the 11 hazards, cyclones have killed more people in India. Second, hazards are seasonal, and so are human deaths. In India, November is cyclone-prone, May is prone to gale and dust storms, April for hailstorms, June for lightning, and January and February for cold snaps. Third, a hazard's effect is spatially determined, so deaths due to disasters are spatially varied. India is diverse, and different regions are exposed to different hazards (Kapur, 2010; Ray-Bennett, 2018).

This perspective provides an excellent insight into the dynamics of geohazards and their effect on humans according to the seasons (Ray-Bennett, 2018). As a result, national and international organisations continue to invest heavily to build the capacity of experts and practitioners by embracing state-of-the-art technologies, such as space technology and multi-hazard early warning systems, to promote effective disaster management systems to reduce deaths (UNISDR, 2015b; Ray-Bennett, 2018). However, this perspective explains little as to why more women and children and persons of other identities die in disasters than men and vice versa or why some groups of people are more vulnerable to disasters than others (Neumayer & Plümper, 2007; Ray-Bennett, 2018). The emerging data suggest that disaster deaths are gendered and classed and occur at the intersection of other social and medical factors, including caste, race, place, the medical causes of death, circumstances, etc. [3]—discussed further in Chaps. 5 and 6.

---

[2] The 13 natural hazards are: cyclone, dust storm, cloudburst, cold wave, drought, dust storm, earthquake, flash flood, gale, hailstorm, heat wave, lightning, snowfall, squall, thunderstorm, and flood (Kapur, 2010).

[3] For instance, WHO (2002) noted that although the Indian Ocean Tsunami 2004 claimed 300,000 lives across 13 nations; 80% of these were women and children. The Indian Meteorological Department found that a large proportion of women (34%) and children (45%) were killed in disasters in comparison to men (21%) between 1977 and 2002 (Kapur, 2010). Neumayer and Plümper's (2007) research of 141 countries between 1981 and 2002 showed that natural disasters lower women's life expectancy more than men. In other words, disasters (and their subsequent impact) on average, kill more women than men and kill women at an earlier age than men. Death rates were almost four times higher amongst socio-economically poor women than men in the 1991 Bangladesh cyclone (Aus Aid, 2014). Women's mortality rates are also much higher in developing countries in comparison to their developed counterparts (Lass et al., 2011; Neumayer & Plümper, 2007; Ray-Bennett, 2018).

Vulnerability Approach

The vulnerability approach aims to explain why some people are more vulnerable to disasters than others, including death during disasters (Ray-Bennett, 2018). According to this approach, the impacts of 'natural disasters' are not entirely 'natural'; instead, they are determined by people's unequal exposure to risks, which are a consequence of socio-economic systems (Cannon, 1994; Neumayer & Plümper, 2007). This approach does not deny the significance of natural hazards as trigger events, but emphasis is placed on the various ways in which social and economic systems can render people more vulnerable to disasters (Cannon, 1994; Ray-Bennett, 2018; Varley, 1994; Winchester, 2000; Wisner et al., 2004). In this approach, the normal daily lives of some vulnerable groups are often difficult to distinguish from disaster conditions (Ray-Bennett, 2018). Disasters only act as an interface between an extreme physical environment and a vulnerable group of population due to a "combination of hazards, conditions of vulnerability and insufficient capacity or measures to reduce the potential negative consequences of risk" (Sivakumar, 2005, p. 2). According to this perspective, differences in mortality to disasters are also explained based on biological, social, cultural, economic, and physical vulnerabilities (Ray-Bennett, 2017, 2018). These vulnerabilities are not distinct; they are often conjoined, intersect, and reinforce each other during times of disaster. A few empirical examples are drawn to demonstrate this.

Biological vulnerability: Physiological differences between men and women put women at high risk during disasters (Ariyabandu, 2009; Neumayer & Plümper, 2007; Ray-Bennett, 2018). Men, in general, are physically stronger than women; therefore, they are likely to withstand the impact of a disaster better than women. For instance, a physically robust man has a better chance to swim or climb up a tree to survive an emerging storm surge. However, biological and physiological differences may also be socially determined (Eklund & Tellier, 2012; Neumayer & Plümper, 2007; Ray-Bennett, 2018).

Social and cultural vulnerabilities: Social and cultural norms related to role behaviour put women, more than men, at greater risk when it comes to rescue efforts (Neumayer & Plümper, 2007; Ray-Bennett, 2018). Dress codes such as *saree* or *burqa* inhibited women's mobility during the 1991 cyclone in Bangladesh. Learning to climb a tree or to swim is socially and culturally not permissible for women and girls in some societies. In Sri Lanka, a study conducted by the Centre on the Epidemiology of Disaster (CRED) noted that only 12–20% of women could swim compared to 75–85% of men (Eklund & Tellier, 2012).

Economic vulnerability: Gender-based division of labour could be equally detrimental to the effect of some hazards (Ray-Bennett, 2018. For instance, more women were killed in the Latur Earthquake in 1993 because, during the time of the earthquake, women and children were at home, whereas men were out working in the field (Kapur, 2010). In contrast, more fishermen and male labourers were killed in the district of Jagatsinghpur in Odisha during the Super-Cyclone 1999 (Samal et al., 2003). The coastal livelihoods in Odisha are reliant heavily on fishing and prawn cultivation. Prawn cultivation is a labour-intensive activity that mainly employs male labourers. When the Super-Cyclone landed, most men were fishing in the sea or rearing the prawns near the coast. Consequently, they were directly on the pathway of the Super-Cyclone (Ray-Bennett, 2009, 2018).

Physical vulnerability: Gender differences can also lead to mortality (Ray-Bennett, 2018). Kapur (2010) argued that greater numbers of women's deaths occurred in the 1993 Latur Earthquake in India due to the nature of house structures. Higher numbers of deaths were recorded in stone and mud houses (86.32%) compared to shacks (0.40%) and brick and mortar (1.15%). It is important to note that not all the poor suffered because the households living in shacks were the least affected. This indicates that poverty and disaster vulnerability are linked but not the same (Cannon, 1994; Ray-Bennett, 2009, 2018).

To summarise, social and cultural vulnerabilities in the interface with other types of vulnerabilities (e.g., sometimes physical, sometimes economic and other times biological, and in some cases as seen above all of these) create sex-based and gender-based vulnerabilities in everyday life and act as a trigger when hazards unfold. According to Kelman (2020a), "The myth of women and girls dying more than men and boys in all disasters needs to be overturned. Instead, different sexes and different genders are forced to be vulnerable in different ways, leading to different disaster death tolls."

Determinants of Disaster Deaths Approach

There is another group of disaster researchers (largely human geographers and epidemiologists) who study the factors or determinants of death using bivariate (e.g. chi-square, *t*-test, and, correlation) and/or multivariate (e.g. multiple regression and its variants) techniques (Paul, 2021). The determinants of death were initially developed to identify causes of infectious diseases (Smith, 1934, quoted in Paul, 2021) under the ambit of the epidemiologic triad or the epidemiologic triangle: Agent (disaster), Host (unit of study), Environment (physical and human). Disaster researchers have used this model to select and examine the determinants of earthquake mortality and injury (Paul, 2021). In the case of disasters triggered by natural hazards, agent factors include the physical characteristics of each type of extreme event (e.g. magnitude, duration, frequency, temporal spacing, spatial extent, spatial dispersion, speed of onset, and diurnal factors) (Paul, 2021). It also includes characteristics such as building materials of residential and non-residential structures, age, height (number of floors), and size of buildings. On the other hand, environmental factors include relevant characteristics of the built and physical environments in which people live. In the context of earthquakes, all these build factors can be associated with the cause of death. Physical environment-related factors include soil types and topographic conduct of the disaster-affected areas. For instance, an area with alluvial soil with high water content may exhibit greater damage, injuries, and deaths from earthquakes than an area with dry and compact soil (Ramirez & Peek-Asa, 2005, quoted in Paul, 2021).

Jonkman and Kelman (2005) integrated this epidemiological triad model with hazard and vulnerability factors to analyse the 'causes and circumstances' of flood deaths in Europe and the United States (Paul, 2021). For the first time, Jonkman and Kelman (2005) draw from event-based data, and the descriptions of the events provide insight into the factors determining the causes of loss of life in floods. Most importantly, they propose a framework to analyse flood-related deaths. This framework includes medical causes of death, loss of life, hazard, and vulnerability factors. All these factors are also known as **'causes and circumstances of death'**.

A combination of hazard factors and vulnerability factors results in flood death due to specific medical causes (Jonkman & Kelman, 2005). For example, rising flood water (hazard) could lead to the evacuation of a retirement home in the flood plain (vulnerability) and two residents having heart attacks (medical cause of death). Medical causes of death are not vulnerability factors in themselves but are the product of the amalgamation of hazard and vulnerability elements (Jonkman & Kelman, 2005). The authors also provide a catalogue to record deaths, which are flood or non-flood, based on the origin and extent of deaths and damage and the resultant societal perception. They also provide a 'classification of flood disaster deaths': medical cause, activity, timing, gender, age, and lack of judgement. These factors contribute to understanding flood-related disaster deaths' causes and circumstances (Jonkman & Kelman, 2005).

Jonkman and Kelman's (2005) model was later modified by Paul (2021). This model classifies disaster-induced deaths into two: hazard and vulnerability factors. The former includes physical characteristics/parameters of the extreme event under consideration (unit of study). Based on the unit of study, the vulnerability factors are also divided into two groups: individual/household factors and factors related to the areal study of the unit. The individual/household factors include demographic and socio-economic factors[4] and the built environment. The factors associated with the areal study of units include median income/GDP per capita, poverty level, urbanisation rate, population density, land area, population size, level of development and political system. The physical environment is also part of this study unit. Although these factors (hazard and vulnerability) are analytically distinct, they do not stand alone in practice. These two factors often simultaneously affect disaster deaths, and as such, they are interconnected (Paul, 2021).

To summarise, these studies aim to understand the 'causes and circumstances' that lead to disaster deaths. These studies add significantly towards the disaster risk reduction perspective. The reason for focusing on the causes of death is also to find the circumstances of death: 'how people die' in a disaster. As Kelman (2020b) puts this succinctly in the context of earthquakes, "It was not really the shaking which

[4] Family income, literacy, age, gender, disability, and house types are some examples of individual/household level vulnerability factors.

killed them, but mainly collapsing infrastructure.[5] [...]. We know exactly what to do, technically and socially, to stop buildings from falling down in earthquakes". In the next chapter, I shall resume discussing the 'cause and circumstances' of deaths in the context of avoidable disaster deaths and how the Government of India and the World Health Organization (WHO) understand them.

Complex Systems Approach

Men, women, children, and persons of other identities' deaths in disasters when attributed to hazards and vulnerability factors and determinants of death cannot be accepted uncritically—other vulnerabilities play a part, too. As such, it can be argued that disaster deaths can occur not only due to the risk and vulnerabilities that exist in the natural, physical (including built-environment) and social systems, as discussed above but also due to vulnerabilities that exist in human-built organisations[6] (Perrow, 1999; Weick, 1990), such as the disaster management system or emergency health system. These systems play a pivotal role in saving lives before, during and after disasters, and their role needs to be amplified to improve disaster responses. This is because actions and inactions before, during, and after a disaster can directly contribute to the circumstances that lead to disaster deaths. I explain this through the lens of 'systems failure' under the ambit of the complex approach (Ray-Bennett, 2017, 2018).

According to this approach, a disaster management system is a conglomeration of different professional groupings and actors designed for specific tasks, functions and goals (Ray-Bennett, 2018). Also, actors[7] or responders within this system adopt different frames of reference. In this vein, I introduced three types of actors or responders in Chap. 1: primary, secondary and tertiary. As such, weak forms of organisation between these actors or responders could potentially lead to *systems failure*, leading to the deaths of vulnerable people in disasters (Ray-Bennett, 2017, 2018). I have introduced a simple compilation of actors and organisations to illustrate my argument in Table 2.1 below in the context of India.

The systems failure has its roots in a complex approach[8]. According to this approach, it is argued that actors and organisations involved in disaster risk management belong to linear systems (Ray-Bennett, 2017, 2018). They are linear because

---

[5] Kelman (2020b) refers to the following earthquakes: "200,000 each in China in 1976 and Haiti in 2010. Farther back in history, perhaps over 800,000 dead in China in 1556; over 200,000 in Syria in 1138; two ninth-century earthquakes in Iran each with six-figure death tolls; and similar tallies twice in Turkey in earlier centuries".

[6] Organisations are a combination of institutions (social design) and a constellation of practices by different actors that give life to the organisation (Wenger, 1998).

[7] Actors are "the agents who carry out or cause to be carried out the main activities of the system, especially its main transformation" (Checkland, 1981, p. 224).

[8] Complex approach has its roots in soft systems thinking (Rittel & Webber, 1973; Bunge, 1977; Midgley, 2014).

Table 2.1 Organisations and actors for COVID-19 and economic lockdown in India

National disaster management authority under the ambit of the Prime Minister	Secondary/Category 2
State Disaster Management Authorities	Secondary/Category 2
Ministry of Home Affairs	Secondary/Category 2
Ministry of Labour and Employment	Secondary/Category 2
Ministry of Health and Family Welfare	Secondary/Category 2
Press Information Bureau	Secondary/Category 2
WHO, UNDRR	Tertiary/Category 3

Source Author

these systems are spatially segregated (Perrow, 1999; Ray-Bennett, 2018). They have dedicated connections and an extensive understanding of risk's nature (Perrow, 1999). They are also highly complex in their structures and processes of work. At the outset, they may look like separate entities, but they are highly interdependent to each other (Ray-Bennett, 2018; Weick, 1990). In this context, disaster deaths are a complex problem (Ray-Bennett, 2018). The complexity arises because the decision to save lives during disasters sits across different governmental departments and organisations (Rittel and Webber, 1973; Ray-Bennett, 2018). I briefly introduced some of the actors and organisations involved in the management and decision-making (***aka* disaster risk governance here**) for COVID-19 and economic lockdown in India in Chap. 1 (also see Table 2.1 above). I shall resume the discussion on disaster risk governance in Chap. 5. To understand how actors and organisations can fail, the systems failure approach provides three interconnected components: coordination failure, communication failure and conflicting world views (Ray-Bennett, 2017, 2018). A brief description of each component is provided below.

Coordination failure: In addition to the coordination of human personnel, the division of labour by function (Prizza, 2007) and stakeholder partnerships, this component focuses on coordinating the flow of the 'core information' (Ray-Bennett, 2018). Core information is early warning information and constitutes near-real-time information about the impact of an event (Comfort et al., 2004). Interpreting this core information involves a complex interaction of humans and technology (Alexander, 1993). Core information is valuable to avoid unnecessary deaths and increase the efficacy of the response system (Comfort et al., 2004). In this context, coordination problems can occur when there is a lack of core information. For example, before the Indian Ocean Tsunami in 2004, there was no early warning system in place to monitor the Indian Ocean water surface. As such, there was no core information for the Category 1 and 2 responders to rely on. The consequence of this was a death toll of more than 225,000 people (Ray-Bennett, 2018; Romo-Murphy et al., 2011).

The decision to coordinate 'core information' can also be hindered by a culture of complacency (Carnegie Foundation expert panel members, Acton & Hibbs, 2012), poor administrative structures (UNISDR, 2006) and a lack of adaptive leadership amidst the convoluted roles and numerous actors present in a DRR climate (multi-stakeholder and multi-sectoral approach) (Aalst et al., 2013; Lalonde, 2011; Prizza, 2007). Furthermore, the coordination problem is closely connected to decision-making (Ray-Bennett, 2018). The core information may be available, but the decision to develop and implement the response system can be hampered by bureaucracy and a lack of political will (Aalst et al., 2013; Alberts & Hayes, 2003; Herrmann, 2009; Ray-Bennett, 2018). Still further, coordination problems can arise under conditions of 'uncertainty' (Kolen & Helsloot, 2012; Ray-Bennet, 2018). Uncertainties and ambiguities are inevitable when dealing with complex problems (Grint, 2008). Some key uncertainties may include assessing potential impacts once the disaster has been predicted, evaluating who should be warned, about what and how (Alexander, 1993); assessing whether to order a mass evacuation, where to, how, when and using what mode of transport (Kolen & Helsloot, 2012); evaluating the health, safety and security of evacuees in emergency shelters (Rashid & Michand, 2000; Ray-Bennett, 2009); and assessing when to cease warnings (Kalsi, 2003; Kolen & Helsloot, 2012).

Communication failure: Communication is *everything* in a disaster management system (Ray-Bennett, 2018). Without communication between the relevant actors and actors to technology, the function of the disaster management system to reduce societal loss and damage from disasters will collapse. This is explained by Ray-Bennett (2018) once again through the flow of the core information generated by early warning systems. The flow of core information can fail due to the physical disruption of early warning systems and the communicating devices, as well as the information's hierarchy. In the information age, telephone, e-mail, satellite phone, mobile, TV, radio, newspaper, paging devices, X (formerly known as Twitter) and the Internet, and these communication devices are central to improving the capacity of the first responders (Comfort et al., 2004), as well as that of the at-risk population (GIZ, 2012; Moore & Verity, 2014; UNISDR, 2006). However, a random failure of the communications networks or power supplies caused by a disaster[9] could significantly damage the flow of information between organisations and the people at risk (Srivastava, 2009; UNDRR, 2022).

Communication can fail in a disaster climate due to rigid hierarchical systems (Ray-Bennett, 2018; Weir, 1996). These overt barriers can manifest due to a system's reliance on traditional chain of command (C) structures where decision-making is centralised and the leader in charge is effectively responsible for disaster management (Ray-Bennett, 2018). Such a system creates a 'mind-set' which disables imagination, dynamism and foresight (Weick, 1990). Rigid hierarchical systems also enable the possibility of core information either evaporating or getting delayed in the structures of human-built organisations (Ray-Bennett, 2018).

Inaccuracy of core information can also lead to systems failure. This inaccuracy can stem from different frames of reference used by professional groups (Weick,

---

[9] On 14 January 2022, the South Pacific experienced the incredible force of the submarine volcano Hunga Tonga-Hunga Ha'apai. Tonga went silent for days. The undersea communications cable connecting Tonga to the rest of the world, and its own islands was severed in two places during the eruption (UNDRR, 2022).

1990; Weir, 1996). Government, non-government and popular media use different forecasting terminology when warning the public and private businesses, which can be confusing (GIZ, 2012, 2015; Herrmann, 2009), and they also differ distinctly from the indigenous early warning systems and practices (GIZ, 2012; Herrmann, 2009; Romo-Murphy et al., 2011). The early warning systems can also fail to provide a reliable warning (Herrmann, 2009). The matter of 'trust' (Morgner, 2013) is of equal importance between the secondary organisations that are generating and cascading the core information and the end users. Also, communication that does not specify the actions required to save lives will be futile. Likewise, a response without core information will be blind in response. These problems can lead to deaths and systems failure in a disaster climate (Ray-Bennett, 2018).

Conflicting World Views: The disaster management system can also fail due to conflicting world views. Under the aegis of the UN's DRR framework, disaster management combines early warning practitioners, Category 1 and 2 responders, and gender and DRR specialists—among others—at international, national and local levels (UN, 2015; Ray-Bennett, 2018). These actors have their world views or *Weltanschauung* on how best to avert disaster risks. Strategies and targets emanate from the world views that these actors adopt. As such, coordination and communication problems are tightly coupled with the conflicting world views of the different actors and organisations of the disaster management system. One of the consequences of these subjective world views is the lack of an 'overall objective' or a 'goal' (Jenkins, 1969) to avoid death. The lack of an objective is instead a lack of input into the system to reduce 'deaths' (Ray-Bennett, 2018). Odisha, an eastern State of India, is a case in point. During Cyclone Phailin (Category 4) in 2014, the Chief Minister of Odisha, Mr Naveen Patnaik set a 'goal' or 'Mission Statement' for the disaster management system, which was 'zero casualty at any cost'. Accordingly, all actors, responders, and organisations organised themselves to achieve this goal, which significantly reduced disaster deaths from 10,000 in 1999 to 86 in 2014 (Ray-Bennett, 2018).

Although more studies are needed to understand the circumstances and causes of disaster deaths (Paul, 2021), the DRR, vulnerability, determinants of disaster deaths and complexity approaches discussed above have been instrumental in providing insights as to why and how deaths occur in the context of disasters triggered by climate-related and geophysical hazards. However, there is currently a paucity of research under the domain of DRR to understand the circumstances and causes of widespread deaths from the impact of COVID-19 and economic lockdown in India or, as a matter of fact, any disasters relevant to lower-middle and low-income countries. Furthermore, these approaches do not go beyond the disaster management system to focus attention on the serious disruption of the functioning of primary and secondary disaster responders and services, including frontline and essential health workers, destruction and disruption of health facilities and services at national and local levels from the impact of disasters leading to deaths. To understand and expand these approaches, I provide an additional layer of analysis by invoking the concept of 'avoidable deaths' from the public health domain. The next chapter extends this concept to 'avoidable disaster deaths'. First, let us focus on what avoidable death is.

Avoidable Deaths Approach

In the public health domain, avoidable death or avoidable mortality has been researched for more than three decades. Yet, there is little consensus among researchers about how to define: what is avoidable death (Olatunde et al., 2016). Despite this, avoidable death is based on the concept that "premature deaths from certain conditions should be rare, and ideally should not occur in the presence of timely and effective health care" according to the Office for National Statistics in the United Kingdom (Olatunde et al., 2016: 4). In the context of this study, I argue that avoidable death is also based on the concept of avoiding premature deaths from certain circumstances including effective *disaster risk governance*—as such, the concept of avoidable death is expanded to include avoidable disaster death—discussed in detail in the next Chap. 3. The Commission of the European Communities Health Services Research defines avoidable deaths as "death from specific diseases (within selected age groups) for which mortality should be wholly or substantially avoidable when appropriate medical care is sought and provided in good time" (Fitzgerald et al., 1993, p. nd).

The concept of avoidable deaths is fluid. Its classification of deaths regarded as avoidable and unavoidable has shifted over the twentieth century into the twenty-first century. Advances in public health, medicines, social policies, and the sophisticated human-built environment have all been crucial in preventing avoidable deaths. Maternal deaths, deaths from tuberculosis, malaria, or other infectious diseases and deaths from animals were unavoidable deaths at some point in time, but they are avoidable now. This duality of avoidability and unavoidability at the interface of time, science, social policy, development, and context is quite a fascinating aspect of the concept of avoidable deaths. However, there are two overlapping trajectories for avoidable deaths. The first focuses on avoidable deaths as a measure of outcome in a *health care set-up,* and the second on *non-health care set-up* for record keeping. Although both are relevant to this study, the latter is emphasised. A brief overview of these trajectories is provided below.

Avoidable Deaths in a Health Care Set-Up

The concept of avoidable deaths was first conceived in 1970 by John Mansfield, a general practitioner in the United Kingdom. According to Mansfield (1970), deaths (especially a vast number of premature deaths) are avoidable in two senses. "Firstly, in the avoidance of factors which are both within man's control and which are harmful. Secondly, in the early detection and treatment of diseases" (Mansfield, 1970, p. 1). Mansfield also identified three main parties who make decisions concerning avoidable deaths. These parties are the individual, the medical professional, and the government. Although the sequence of events leading up to avoidable disease, death, or disability of an individual can be complex, they are identifiable.

For example, a case of diphtheria may be a consequence of failure by: a government to introduce the necessary legislation to promote or fund immunization programmes; health service managers to implement such programmes; a doctor to immunize patients; parents to take their baby to a clinic because it is too far away or because of the family's religious views' (Fitzgerald et al., 1993, p. nd).

The concept of avoidable deaths identifies who is responsible for avoiding deaths and a pathway for preventing them. This level of clarity is also essential in disaster response and recovery to promote accountability and capacity of the disaster risk management systems at local and national levels.

The concept of avoidable deaths was then popularised by the Preventable and Manageable Diseases Working Group led by Rutstein et al., (1976) at Harvard University in the United States. David Rutstein and his colleagues developed outcome measures for health services by counting cases of 'unnecessary disease and disability, and unnecessary untimely deaths' as measures of the quality of medical care. They argued that to develop effective indicators of healthcare, a number of disease lists should be drawn up, which should not, or should only infrequently, give rise to death or disability (Rutstein et al., 1976; quoted in Olatunde et al., 2016). Rutstein also noted that the list of conditions considered to be avoidable would need to be updated in light of improvements in medical knowledge and practice, as well as social and environmental changes (Rutstein et al., 1976). According to them, advancements in science and technology, coupled with the ever-changing 'social scene', would require these priorities to be updated over time.

Rutstein et al. (1983) also established quantitative negative indexes of health through which cases of (i) unnecessary disease, (ii) unnecessary disability, and (iii) unnecessary untimely deaths could be counted. Negative health indexes, such as maternal and infant mortality rates, have already successfully saved many lives worldwide (e.g. Millennium Development Goal 5, currently the Sustainable Development Goal 3). However, as they apply only to mothers and infants, there was a need to add negative indexes for all unnecessary diseases, disabilities, and untimely deaths (Rutstein et al., 1976). They developed a list of over 90 conditions to be used in local and national geographic areas and individual hospitals. In selecting these conditions as indexes of the quality of care, it was assumed that "if everything had gone well, the condition could have been prevented or managed" (Rutstein et al., 1983, p. 583). This list of conditions was designed for international use and included even those which did not occur in the United States, such as cholera, but were crucial international health conditions. Wherever possible, the conditions were selected from the 8th revision of the WHO's International Classification of Diseases (ICD) adapted for use in the United States to ensure ease of comparison with existing health data in the United States (Rutstein et al., 1983).

Charlton et al., (1983) were the first to apply the concept of avoidable deaths empirically at a population level to examine trends at the national and global level, the significance of social factors, and the incidence of disease. Their study was focused on mortality data from England and Wales for the period of 1974 to 1978. Mortality was selected for the study in place of morbidity or disability, as data on the former is more reliable regarding availability, frequency, geographical representation, accuracy, and completeness than the latter two. The mortality reported was influenced by factors such as environmental, genetic, nutritional, and social factors, as well as local variations in diagnostic habits and precision. The mortality indicators they suggested were intended to shed light on where problems may lie and whether to initiate further inquiry to provide a definitive evaluation. While Rutstein et al.'s., list of conditions included over 90 conditions from which deaths were considered avoidable through medical interventions or unnecessary untimely mortality, Charlton et al., (1983) selected 14 causes of untimely death among the 98 area health authorities (AHAs) of England and Wales. These authors emphasised that it would be unwise to judge the performance of Area Health Authorities based only on the reported 'avoidable mortality' as these were intended only as "warning signals of possible shortcomings" in health care delivery (Charlton et al., 1983, p. 696). They hoped that monitoring cases of avoidable mortality would eventually lead to reduced death rates.

Following Charlton et al.'s., publications, a working group within the European Community, the Concerted Action Project on Health Services, extended the concept of avoidable deaths by collecting data on fatalities caused by preventable and amenable conditions. This data was presented as an Atlas, illustrating 5-10-fold intra- and international variations in mortality in Europe (Holland, 1986). The European Community defined 'health care' as primary care, hospital care, and collective health services. Although it is challenging to attribute high mortality rates to one cause with absolute certainty, Holland (1986, p. 117) posits that the avoidable deaths indicators put forward by the Commission of the European Community, which covered a range of health services, were intended to signal possible shortcomings in the delivery of health care services and not "designed to point the finger at poorly serviced health areas".

Another study on avoidable deaths by Poikolainen and Eskola (1988) explored whether investments in health care were worthwhile and had the intended impact on health and mortality by studying the relationship between mortality and health service resources, which included the number of medical doctors, nurses, and hospital beds available. Poikolainen and Eskola (1988, p. 87) proposed four significant categories of mortality vis-à-vis gross domestic product: (i) deaths amenable to intervention by health services (later termed as "amenable deaths"); (ii) deaths not amenable to such an intervention (later termed "non-amenable deaths"); (iii) other deaths from natural causes (later termed "partly-amenable deaths"); and (iv) deaths from violent causes. The authors found a strong negative correlation between: (i) gross domestic product and mortality from non-amenable causes among males; (ii) gross domestic product and mortality resulting from partly-amenable causes among females; and (iii) the number of hospital beds and mortality resulting from partly amenable causes among females. However, a positive correlation was found between: (i) the number of nurses and midwives and mortality from non-amenable and partly amenable causes among females; (ii) alcohol consumption and non-amenable causes among males (Poikolainen & Eskola, 1988, p. 87).

It is emphasised that general socio-economic development, as indicated by a higher gross domestic product, does not assure lower mortality rates from amenable causes. Further, it cannot be said that investments in health care are irrelevant. Instead, the observed high positive correlations between nations' gross domestic product and the number of available nurses and hospital beds suggest that affluent nations have responded to the health sector's requirements by increasing the availability of services where they are most needed. This may also explain the positive correlation observed between the number of doctors and mortality found in young age groups (Poikolainen & Eskola, 1988).

Mackenbach et al. (1988, p. 889) analysed trends in mortality from a selection of conditions suggested by Rutstein et al.'s., lists of 'unnecessary untimely mortality' to evaluate the impact of medical innovations in the Netherlands post-1950. Mackenbach et al. (1988) hold that reductions after 1950 in mortality in conditions which have become amenable to adequate medical care, at least to a degree, have contributed significantly to the increases in life expectancy observed. However, a decline in mortality from these conditions cannot be solely attributed to medical care innovations.

Studies on avoidable deaths started shifting after Mackenbach et al. (1988) by focusing "on differentiating and comparing levels of 'avoidable' mortality attributable to the health care system and wider health policies usually not within the direct control of health services" (Kossarova et al., 2009, p. 8). These authors were the first to compare preventable and treatable conditions explicitly (Kossarova et al., 2009), where the avoidable deaths indicators were "divided into those indicating the outcome of health care intervention and those mainly indicating the results of national health policies" (Westerling & Smedby, 1992, p. 503). They are credited with introducing the term 'medical care indicators', which refers specifically to medical conditions amenable to medical care, as opposed to 'health policy indicators' referring to conditions responsive to primary prevention measures (wider health policies) (Nolte & McKee, 2004).

Carrying forward Westerling and Smedby's work on avoidable deaths, Simonato et al. (1998) analysed the trends of avoidable mortality in 21 countries of Europe between 1955 and 1994. These countries were divided into quadrants, namely northern, central and southern Europe, as well as Nordic countries. Their study made an essential contribution to the study of avoidable mortality through a categorisation of causes of death by level of prevention into the following three groups, which are:

(i) Preventable through primary prevention, that is, by reducing exposure as well as the incidence of disease. It includes conditions caused by lifestyle factors (such as tobacco and alcohol use); other conditions principally connected to the use of alcohol; deaths from injury and poisoning (such as traffic safety and crime).
(ii) Amenable to secondary prevention through early detection and treatment.
(iii) Amenable to tertiary prevention through improved treatment and medical care (such as mortality from infectious diseases, appendicitis, and hernias, among others) (Simonato et al., 1998).

Tobias and Jackson (2001) studied avoidable deaths in New Zealand between 1981 and 1997 to assess the capacity of the nation's population's capacity to improve health outcomes. The three subcategories of avoidable deaths that were used are as follows, which were then weighted for each potentially avoidable condition: (i) Primary Avoidable Mortality (PAM): This group includes conditions that can be prevented by individual behavioural changes or public health interventions; (ii) Secondary Avoidable Mortality (SAM): This group includes conditions that are amenable to early detection and intervention, generally in the context of primary health care; and (iii) Tertiary Avoidable Mortality (TAM): This group includes conditions whose fatality rates can be significantly reduced through existing medical or surgical procedures, even in cases where the disease process is completely developed. Tobias and Jackson's study revealed great scope for improvement within the health sector and for social policies and services to contribute to overall population health gain and improve the equitable nature of health outcomes across various socio-economic and ethnic groups.

Although the above literature does not focus on avoidable deaths from disasters or epidemics or on lower-middle and low-income countries, it provides insight into the evolving trajectory of the concept and practice of avoidable deaths as a measurement in hospitals and health care systems in the context of a nation's economy and a wider health policy set-up.

Avoidable Deaths in a Non-health Care Set-Up

The Office for National Statistics (ONS) in the United Kingdom provides additional understanding of the concept of avoidable deaths in the context of non-health care set-up[10]. Annually the Office for the National Statistics publishes an online report on '*Avoidable Mortality in the UK*'. Figures for the United Kingdom are calculated using death registration data for England and Wales held by the Office for National Statistics and death registration data for Scotland and Northern Ireland provided by the National Records of Scotland and the Northern Ireland Statistics and Research Agency, respectively (Price, 2021).

According to the Office for National Statistics, avoidable deaths are "deaths that are amenable, preventable or both, where each death is counted only once" (Price, 2021, p. 2). Amenable or treatable mortality are "deaths that could be avoided through timely and effective healthcare, and preventable mortality or deaths that could be avoided by public health interventions" (Price, 2021, p. 2). However, these terms are not discussed at length by the Office of National Statistics (see Price, 2021; Cook, 2019; Olatunde et al., 2016), as such, I have reviewed additional literature in the next chapter. These terminologies are important to this study in the absence of literature on avoidable deaths in India and shall act as a reference in the later chapters.

---

[10] "The avoidable deaths in hospital measure is based on a record review of a sample of deaths deemed to be due to problems in care. Avoidable deaths in hospital data are not intended to be comparable and are not currently collated centrally" in the United Kingdom (Price, 2021, p. 12).

Since my working definition of a disaster emphasises the "serious disruption of the functioning of primary and secondary disaster responders and services, destruction and disruption of health facilities and services [...]" (Chap. 1), the concept of avoidable death discussed above is insightful in understanding deaths by causes and circumstances in the context of disasters and emergencies. The causes and conditions, translated using the WHO's International Classification of Diseases Tenth Revision (currently ICD-10), are not straightforward in relating to causes and conditions. As such, Olatunde et al. (2016) suggest a *pragmatic approach* to identifying avoidable deaths. In this light, I adopted a *pragmatic approach* to understanding disaster deaths in COVID-19 and economic lockdown by considering *disaster risk governance* integral to the two essential components of avoidable deaths advocated by the Office for National Statistics: *amenable* and *preventable*. Together, they form the avoidable disaster deaths (ADD) approach.

Conclusion

At the outset of this chapter, I raised five questions: Why do people die in disasters? Who are those people? What are the causes and circumstances that lead to deaths? And how can we reduce disaster deaths? I attempted to answer these questions through the state-of-the-art literature on disaster risk reduction, vulnerability, determinants of disaster-induced deaths, systems failure, and avoidable deaths. Reducing disaster deaths or avoidable disaster deaths is a complex problem, and the engagement with different approaches demonstrates its complexity as well as solutions to reduce them. My attempt to answer these questions continues to the next chapter, where I engage with the concepts of avoidable deaths and disaster deaths once again, and I coin this discourse as 'avoidable disaster deaths'.

References

Aalst, M. V., Kellett, J., Pichon, F., & Mitchell, T. (2013). *Incentives in disaster risk management and humanitarian response: Background note for world development report*. Overseas Development Institute.

Acton, J. M., & Hibbs, M. (2012). *Why Fukushima was preventable*. The Carnegie Papers. http://carnegieendowment.org/files/fukushima.pdf. Accessed on December 23, 2023.

Alberts, D. S., & Hayes, R. E. (2003). *Power to the edge: Command, control in the information age*. CCRP Publication Series. Available at: www.dodccrp.org. Accessed on December 23, 2023.

Alexander, D. (1993). *Natural disasters*. UCL Press.

Ariyabandu, M. M. (2009). Sex, gender and gender relations in disasters. In E. Enarson, & P. G. D. Chakrabarti (Eds.), *Women, gender and disaster: Global issues and initiatives*. Sage Publications.

Ariyabandu, M. M., & Wicramasinghe, M. (2003). *Gender dimensions in disaster management: A guide for South Asia*. ITDG South Asia Publication.

Aus Aid. (2014). Women's and children's health in emergencies. Aus Aid. Available at: https://www.dfat.gov.au/aboutus/publications/Pages/australia-s-humanitarian-aid-women-s-and-children-s-health-in-emergencies. Accessed on February 11, 2025.

Bryant, E. (2005). *Natural hazards* (2nd ed.). Cambridge University Press.

Bunge, M. (1977). General systems and holism. *General Systems, 22*, 87–90

Cannon, T. (1994). Vulnerability analysis and the explanation of "natural disasters". In A. Varley (Ed.), *Disasters, development and environment*. Wiley.

Charlton, J. R. H., Silver, R., Hartley, R. M., & Holland, W. W. (1983). Geographical variation in mortality from conditions amenable to medical intervention in England and Wales. *The Lancet, 321*(8326), 691–696. Available at: https://www.sciencedirect.com/science/article/pii/S0140673683919815. Accessed on October 22, 2020.

Checkland, P. (1981). Systems thinking: Systems practice. Wiley, Chichester.

Comfort, L. K., Ko, K., & Zagoreck, A. (2004). Coordination in rapidly evolving disaster response systems: The role of information. *American Behavioural Scientist, 48*(3), 295–313. https://doi.org/10.1177/0002764204268987

Cook, A. (2019). Avoidable mortality in the UK: 2017. In *Statistical bulletin*. Office of National Statistics. Available at: https://www.ons.gov.uk/peoplepopulationandcommunity/healthandsocialcare/causesofdeath/bulletins/avoidablemortalityinenglandandwales/2017. Accessed on April 19, 2022.

Davis, I., & Gupta, S. P. (1991). Technical background paper. In Asian Development Bank (Ed.), *Disaster mitigation: In Asia and the Pacific*. Asian Development Bank.

Eklund, L., & Tellier, S. (2012). Gender and international crises response: Do we have the data, and does it matter? *Disasters, 36*(4), 589–608. https://doi.org/10.1111/j.1467-7717.2012.01276.x

Fitzgerald, A., et al. (1993). *European community atlas of 'avoidable death'*. Commission of the European Communities Health Services Research Series No. 9. Oxford University Press.

GIZ (Deutsche Gesellschaft fur Internationale Zusammenarbeit). (2012). *LFEWS: Local flood and early warning system*. GIZ Publication.

GIZ (Deutsche Gesellschaft fur Internationale Zusammenarbeit). (2015). *Harmonisation of early warning alert levels*. GIZ Publication.

Grint, K. (2008). Wicked problems and clumsy solutions: The role of leadership. *Clinical Leader, 1*(II), 11–15.

Herrmann, S. (2009). Remote sensing. In M. Glantz (Ed.), *Heads up!: Early warning systems for climate, water and weather-related hazards*. United Nations University Press.

Holland, W. (1986). The 'avoidable death' guide to Europe. *Health Policy, 6*, 115–117. Available at: https://www.sciencedirect.com/science/article/abs/pii/0168851086900011. Accessed on October 28, 2020.

Jenkins, G. M. (1969). The systems approach. *Journal of Systems Engineering, 1*, 3–49.

Jonkman, S. N., & Kelman, I. (2005). An analysis of causes and circumstances of flood disaster deaths. *Disasters, 29*(1), 75–97.

Kalsi, S. R. (2003). Monitoring and forecasting of tropical cyclones and their associated effects. In H. K. Gupta (Ed.), *Disaster management*. Universities Press.

Kapur, A. (2010). *Vulnerable India: A geographical study of disasters*. Sage Publications.

Kelman, I. (2020a). Do disasters kill more women than men? *Psychology Today*, 2 December. Available at: https://www.psychologytoday.com/gb/blog/disaster-choice/202012/do-dsasters-kill-more-women-men. Accessed on January 16, 2023.

Kelman, I. (2020b). Dying in earthquakes: A tale of umpteen tremors. *Psychology Today*, 10 August. Available at: https://www.psychologytoday.com/gb/blog/disaster-choice/202008/dying-in-earthquakes-tale-umpteen-tremors. Accessed on January 16, 2023.

Kolen, B., & Helsloot, I. (2012). Time needed to evacuate the Netherlands in the event of large scale flooding: Strategies and consequences. *Disasters, 36*(4), 700–722. https://doi.org/10.1111/j.1467-7717.2012.01278.x

Kossarova, L., Holland, W., Nolte, E., & McKee, M. (2009). Measuring 'Avoidable' mortality: Methodological note. In *Directorate-general for employment, social affairs and equal opportunities*, Brussels, Belgium. Available at: http://eprints.lse.ac.uk/46390/1/Measuring%20avoidable%20mortality%20(lsero).pdf. Accessed on October 16, 2020.

Lalonde, C. (2011). Managing crisis through organisational development: A conceptual framework. *Disasters, 35*(2), 443–464. https://doi.org/10.1111/j.1467-7717.2010.01223.x

Lass, W., Haas, A., Hinkel, J., & Jaeger, C. (2011). Avoiding the avoidable: Towards a European heat waves risk governance. *International Journal of Disaster Risk Science 2*(1), 1–14. https://doi.org/10.1007/s13753-011-0001-z

Mackenbach, J. P., Looman, C. W., Kunst, A. E., Habbema, J. D. F., & Van Der Maas, P. J. (1988). Post-1950 mortality trends and medical care: Gains in life expectancy due to declines in mortality from conditions amenable to medical intervention in The Netherlands. *Social Science and Medicine, 27*(9), 889–894. Available at: https://www.sciencedirect.com/science/article/pii/027795368890278X. Accessed on October 28, 2020.

Mansfield, J. (1970). *Avoidable death: Survival and commonsense.* Cassel and Company Ltd.

Midgley, G. (2014). *An introduction to systems thinking for tackling wicked problems.* M.Sc. in Risk, Crisis and Disaster Management Study School, University of Leicester, 29 September.

Moore, R., & Verity, A. (2014). *Hashtag standards for emergencies: OCHA policy and studies brief*. United Nations Office for the Coordination of Humanitarian Affairs (OCHA).

Morgner, C. (2013). Trust and confidence: History, theory and socio-political implications. *Human Studies, 36*(4), 509–532. https://doi.org/10.1007/s10746-013-9281-1

Neumayer, E., & Plümper, T. (2007). The gendered nature of natural disasters: The impact of catastrophic events on the gender gap in life expectancy, 1981–2002. *Annals of the Association of American Geographers, 97*(3), 551–566. https://doi.org/10.1111/j.1467-8306.2007.00563.x

Nolte, E., & McKee, M. (2004). *Does health care save lives? Avoidable mortality revisited.* The Nuffield Trust. Available at: https://drum.lib.umd.edu/bitstream/handle/1903/24477/does-health care-save-lives-mar04-1.pdf?sequence=1. Accessed on November 4, 2020.

Olatunde, O., Windsor-Shellard, B., & Campbell, A. (2016). *Revised definition of avoidable mortality.* Office for National Statistics. Available at: www.ons.gov.uk. Accessed on September 27, 2023.

Paul, B. K. (2021). *Disaster deaths: Trends, causes and determinants.* Routledge.

Perrow, C. (1999). *Normal accidents: Living with high-risk technologies* (2nd ed.). Princeton.

Poikolainen, K., & Eskola, J. (1988). Health services resources and their relation to mortality from causes amenable to health care intervention: A cross-national study. *International Journal of Epidemiology, 17*(1), 86–89. Available at: https://academic.oup.com/ije/article-abstract/17/1/86/750265. Accessed on October 26, 2020.

Price, M. (2021). Avoidable mortality in the UK: 2019. In: *Statistical bulletin.* Office for National Statistics. Available at: https://www.ons.gov.uk/peoplepopulationandcommunity/healthandsocialcare/causesofdeath/bulletins/avoidablemortalityinenglandandwales/2019. Accessed on September 27, 2023.

Prizza, R. (2007). The role of coordination in disaster management. In J. Pinkowski (Ed.), *Disaster management handbook.* CRC Press.

Rashid, F. S., & Michand, S. (2000). Female adolescents and their sexuality: Notions of honour, shame, purity and pollution during the floods. *Disasters, 24*(1), 54–70. https://doi.org/10.1111/1467-7717.00131

Ray-Bennett, N. S. (2009). *Caste, class and gender in multiple disasters: The experiences of women-headed households in an Oriya village.* VDM Verlag.

Ray-Bennett, N. S. (2017). Disasters, deaths and the Sendai goal one: Lessons from Odisha, India. *World Development, 103*(2018), 27–39. https://doi.org/10.1016/j.worlddev.2017.10.003

Ray-Bennett, N. S. (2018). *Avoidable deaths: A systems failure approach to disaster risk management.* Springer Nature.

Ray-Bennett, N. S., Mendez, D., Alam, E., & Morgner, C. (2020). Inter-agency collaboration in natural hazard management in developed countries. In B. Gerber (Ed.), *Oxford research encyclopedia of natural hazard science*. Oxford University Press. https://doi.org/10.1093/acrefore/9780199389407.013.176

Rittel, H. W. J., & Webber, M. M. (1973). Planning problems are wicked problems. In: Cross, N. (Ed.), *Developments in design methodologies*. Wiley (*Policy Science, 4*, 155–169).

Romo-Murphy, E., James, R., & Adams, M. (2011). Facilitating disaster preparedness through local radio broadcasting. *Disasters, 35*(4), 801–815. https://doi.org/10.1111/j.1467-7717.2011.01234.x

Rutstein, D. D., Berenberg, W., Chalmers, T. C., Child, C. G., 3rd., Fishman, A. P., Perrin, E. B., Feldman, J. J., Leaverton, P. E., Lane, J. M., Sencer, D. J., & Evans, C. C. (1976). Measuring the quality of medical care: A clinical method. *New England Journal of Medicine, 294*(11), 582–588.

Rutstein, D. D., Mullan, R. J., Frazier, T. M., Halperin, W. E., Melius, J. M., & Sestito, J. P. (1983). Sentinel health events (occupational): A basis for physician recognition and public health surveillance. *American Journal of Public Health, 73*(9), 1054–1062. Available at: https://www.ncbi.nlm.nih.gov/pmc/articles/PMC1651048/?page=1. Accessed on October 16, 2020.

Samal, K. C., Meher, S., & Panigrahi, N. (2003). *Beyond relief disaster mitigation, livelihood rehabilitation and the post-cyclone recovery in Orissa: Village level studies in three most cyclone affected districts in Orissa*. Nabo Krishna Centre for Development Studies Publication.

Simonato, L., Ballard, T., Bellini, P., & Winkelmann, R. (1998). Avoidable mortality in Europe 1955–1994: A plea for prevention. *Journal of Epidemiology & Community Health, 52*(10), 624–630. Available at: https://jech.bmj.com/content/jech/52/10/624.full.pdf

Sivakumar, M. V. K. (2005). Impacts of natural disasters in agriculture, rangeland and forestry: An overview. In M. V. K. Sivakumar, R. P. Motha, & H. P. Das (Eds.), *Natural disasters and extreme events in agriculture: Impacts and mitigation*. Springer.

Srivastava, S. K. (2009). Making a technological choice for disaster management and poverty alleviation. *Disasters, 33*(1), 58–81. https://doi.org/10.1111/j.1467-7717.2008.01062.x

Tobias, M., & Jackson, G. (2001). Avoidable mortality in New Zealand, 1981–97. *Australian and New Zealand Journal of Public Health, 25*(1), 12–20. Available at: https://doi.org/10.1111/j.1467-842X.2001.tb00543.x. Accessed on October 29, 2020.

UN (United Nations). (2015). *Sendai framework for disaster risk reduction 2015–2030*. UNISDR. https://www.unisdr.org/files/43291_sendaiframeworkfordrren.pdf. Accessed on June 24, 2016.

UNDRR (United Nations Office for Disaster Risk Reduction). (2022). *The invisible toll of disasters*. UNDRR. https://www.undrr.org/explainer/the-invisible-toll-of-disasters-2022. Accessed on October 2, 2024.

UNISDR (United Nations International Strategy for Disaster Reduction). (2005). Hyogo framework for action 2005–2015: Building the resilience of nations and communities to disasters. In *World Conference on Disaster Reduction, Kobe, Hyogo, Japan*, January 18–22, 2005. Available at: https://www.unisdr.org/2005/wcdr/intergover/official-doc/L-docs/Hyogo-framework-for-action-english.pdf. Accessed October 13, 2024.

UNISDR (United Nations International Strategy for Disaster Reduction). (2006). Developing early warning systems, a checklist. In *Third International Conference on Early Warning*. http://www.unisdr.org/we/inform/publications/608. Accessed on January 14, 2014.

UNISDR (United Nations International Strategy for Disaster Reduction). (2015a). *Making development sustainable: The future of disaster risk management: Global assessment report on disaster risk reduction 2015*. UNISDR.

UNISDR (United Nations International Strategy for Disaster Reduction). (2015b). Japan provides early warning example. In *UN World Conference in Disaster Risk Reduction 2015 Sendai Japan. Third UN World Conference on Disaster Risk Reduction*. Available at: www.wcdrr.org. Accessed on March 16, 2015.

UNISDR (United Nations International Strategy for Disaster Reduction) (2018). *2018 annual report*. United Nations.

Varley, A. (1994). The exceptional and the everyday: Vulnerability analysis in the international decade for natural disasters reduction. In A. Varley (Ed.), *Disasters, development and environment*. Wiley.
Weick, K. E. (1990). The vulnerable systems: An analysis of the Tenerife air disaster. *Journal of Management, 16*(3), 571–593. https://doi.org/10.1177/014920639001600304
Weir, D. T. H. (1996). Risk and disasters: The role of communications breakdown in plane crashes and business failure. In C. Hood, & C. Jones (Eds.), *Accident and design: Contemporary debates in risk management*. UCL Press.
Wenger, E. (1998). *Communities of practices: Learning, meaning, and identity*. Cambridge University Press, New York.
Westerling, R., & Smedby, B. (1992). The European Community 'Avoidable Death Indicators' in Sweden 1974–1985. *International Journal of Epidemiology, 21*(3), 502–510. Available at: https://academic.oup.com/ije/article-abstract/21/3/502/695036. Accessed on October 29, 2020.
WHO (World Health Organization). (2002). Gender and health in natural disasters. WHO: gender, equity, human rights. Available at: https://iris.who.int/handle/10665/68886. Accessed on February 11, 2025.
Winchester, P. (2000). Cyclone mitigation, resource allocation and post-disaster reconstruction in south India: Lessons learnt from two decades of research. *Disasters, 24*(2), 18–37.
Wisner, B., Blaikie, P., Cannon, T., & Davis, I. (2004). *At risk: Natural hazards, people's vulnerability and disasters* (2nd ed.). Routledge.
WMO (World Meteorological Organization). (2021). *WMO atlas of mortality and economic losses from weather, climate and water extremes (1970–2019)*. WMO Publication. Available online: https://library.wmo.int/viewer/57564/?offset=#page=1&viewer=picture&o=bookmarks&n=0&q=
Zaman, M. Q. (1999). Vulnerability, disaster, and survival in Bangladesh: Three case studies. In: O. Smith, & S. Hoffman, (Eds.), *The angry earth: Disaster in anthropological perspective*. Routledge.

Chapter 3
How Can Disaster Deaths Be Avoided?

> *Tackling a social calamity is not like fighting a war [...]. in contrast, What is needed for dealing with a social calamity is participatory governance and alert public discussion.*
> —Sen (2020)

Abstract This chapter introduces the 'avoidable disaster deaths' (ADD) approach. The concept of 'unavoidable deaths' is also discussed. The concept of 'avoidable deaths' has its roots in public health and epidemiology. This study has expanded this concept to 'avoidable disaster deaths' by invoking theories and approaches from disasters, systems thinking, geography, crisis and risk management. By doing so, this approach is distinct from 'avoidable deaths'. This chapter also introduces templates and matrices for policymakers and practitioners to make sense of avoidable disaster deaths and unavoidable deaths so that the capacity of the disaster response system can be developed to save lives in a disaster climate.

Keywords Avoidable disaster deaths · Disaster deaths · Avoidable deaths · Unavoidable deaths

Introduction

The previous chapter addressed several questions: Why do deaths occur in disasters? Who dies? What are the causes and circumstances surrounding those deaths? How can we reduce disaster deaths? This chapter continues the focus on the last question by first explaining what 'disaster deaths' are and, second, how we can reduce them. To reduce disaster deaths, I introduce a novel approach of 'avoidable disaster deaths' to make sense of disaster-related deaths in the first phase of COVID-19 and the economic lockdown in India. This sense-making tool is applicable in other disaster contexts, too. I introduce this approach in relation to the existing state-of-the-art literature on disaster deaths, avoidable deaths, determinants of disaster-induced deaths, and systems failure (see Chap. 2). In this chapter, I engage in-depth with the concept of

N. S. Ray-Bennett, *Avoiding Disaster Deaths*,
https://doi.org/10.1007/978-3-031-81300-9_3

disaster deaths advocated by the United Nations Office for Disaster Risk Reduction (UNDRR) before presenting the 'avoidable disaster deaths' approach.

Disaster Deaths Approach

Disaster deaths *aka* disaster mortality/fatality in the context of disaster risk reduction were conceived in the Sendai Framework in 2015. The Sendai Framework set seven global Targets and 38 indicators for national governments to achieve by 2030. Of the seven Targets, the first two targets relevant to this study were introduced in Chap. 1: "Substantially reduce global disaster mortality by 2030" (Target A) and "Substantially reduce the number of affected people globally by 2030" (Target B) (UN, 2015, p. 9).

Although the impact of disasters is loss and damage, there was no direct focus on reducing 'disaster deaths' or human loss in the 'Hyogo Framework for Action 2005–2015' (Ray-Bennett, 2018). Due to the first two global Targets of the Sendai Framework, reducing disaster deaths is currently a high priority for the UN Member States (UN, 2015). According to the Sendai Framework 'disaster deaths' are viewed through the lens of 'reducing global disaster mortality by 2030' (Target A). This logic has led to a focus on tracking the Target through disaster death stocktaking and monitoring systems and developing indicators to improve methodology and auditing of deaths (discussed below). As such, the concept of 'disaster deaths' for Target A has not immediately led to a focus on *reducing* these deaths. The 'avoidable disaster deaths' approach is proposed to overcome this gap. First, I focus on how 'reducing global disaster mortality by 2030' is advocated and implemented by the UNDRR and the UN Member States.

Stocktaking

Two years after the launch of the Sendai Framework, the United Nations General Assembly established the Expert Working Group in 2017 to develop the indicators, terminology, and technical guidance essential for achieving the seven global Targets (Ray-Bennett et al., 2022; UN General Assembly, 2016; UNISDR, 2017). The technical guidance provided by the Expert Working Group is about setting a minimum standard for the UN Member States to record disaster deaths. Although the Sendai Framework does not dedicate a specific section to the implementation details for Target A, it is assumed that this Target can be achieved through the implementation of the four Priorities for Action (see Table 3.1) and in collaboration with stakeholders (section V) and international cooperation and global partnership (section VI) (see UN, 2015).

Table 3.1 Four priorities for action

Priority for action 1	Understanding disaster risk
Priority for action 2	Strengthening disaster risk governance to manage disaster risk
Priority for action 3	Investing in disaster risk reduction for resilience
Priority for action 4	Enhancing disaster preparedness for effective response and to 'Build Back Better' in recovery, rehabilitation and reconstruction

Source UN (2015)

The seven Global Targets are currently monitored through the DesInventar Open Source[1] and Sendai Monitor initiated by the UNDRR (Mizutori, 2020). While writing this chapter in February 2023, according to the Sendai Monitor,[2] of the 195 countries that have signed up for the Sendai Framework, 76 and 81 countries are reporting for Target A and Target B, respectively. This means 119 countries for Target A and 114 for Target B are not reporting. Although the Sendai Monitor is a step change, the reporting of the Sendai Monitor indicates that the uptake of this tool is low amongst the UN Member States. This low uptake, however, does not mean that the national governments are not recording disaster deaths. It is likely that the indicators and terminologies suggested by the Expert Working Group for disaster deaths are either not used or inconsistently used at national levels (Bahinipati & Gupta, 2022; Gupta et al., 2022).

Indicators and Terminologies

The Expert Working Group developed the indicators and terminologies for Target A (UNISDR, 2017). The Inter-Agency and Expert Working Group on Sustainable Development Goal Indicators (IAEG-SDGs) proposed the use of these same indicators in measuring disaster-related global Targets of SDGs 1 (No Poverty), 11 (Sustainable Cities and Communities) and 13 (Climate Action) (UNISDR, 2017). Doing so reinforces the importance of the Sendai Framework's Target A and indicators (Ray-Bennett et al., 2022).

The Expert Working Group's technical guidance defines the "data collection that Member States should consider in order to develop a robust methodology to measure mortality" (UNISDR, 2017, p. 6). They suggest that the data collection on Target A should focus on (i) location; (ii) disaggregation by disability; (iii) attribution to a disaster; (iv) the hazard associated with a disaster; (v) temporal aspects for attribution and cut-off for data collection; (vi) biological hazards; (vii) set of hazards, and (viii)

[1] https://www.desinventar.net/

[2] https://sendaimonitor.undrr.org/; https://sendaimonitor.undrr.org/analytics/global-target/13/6.

countries starting loss data collection and are yet to establish a clear legal framework for these criteria are recommended to adopt the following approach: hazard; cause of death; time–space of the recommended cut-off period; sources of data (for details, see UNISDR, 2017, pp. 10–12). Although these data collection variables do not fully align with the 'causes and circumstances' variables introduced in the previous chapter (reference to Jonkman & Kelman, 2005; Paul, 2021), nonetheless, the UNDRR's variables are meant to act as 'minimum standards' to capture the 'causes and circumstances' surrounding a disaster death at a national level.

Targets and terminology are essential for operational clarity for national and international organisations, and these were missing from the Hyogo Framework (Ray-Bennett, 2017, 2018; Ray-Bennett et al., 2022). As such, this technical guidance to UN Member States is a significant improvement in measuring progress toward global targets across countries and over 15 years of the Sendai Framework and SDGs by sharing minimum standards (common and detailed international understanding of indicators, data required, and providing standard methodologies) for countries which may want to use them voluntarily (UNISDR, 2017).

Despite this step-change, it is not yet clear if the national governments have adopted these indicators and definitions in all low, middle, and high-income countries (Ray-Bennett et al., 2022). Based on the data available (as mentioned above) on the Sendai Monitor platform (as of February 2023), about 119 countries for Target A and 114 for Target B were not reported. A recent study conducted by Bahinipati and Gupta (2022) discussed the challenges related to a lack of clarity regarding the existing definitions and methods associated with estimating not just economic loss and damage (ELD) but also non-economic loss and damage (NELD) in India (with implications for other lower-middle and low-income countries) from extreme events and slow-onset disasters, particularly for cyclonic storms, floods, and droughts. Nevertheless, the Expert Working Group's indicators are inspiring, clearly understood, few in numbers, ambitious yet feasible, and, importantly, measurable (Maini et al., 2017).

To understand the terminology of 'death' or 'disaster death' in detail, I invoke the Integrated Research on Disaster Risk's (IRDR, 2015) *Guidelines on Measuring Losses from Disasters.* This Guideline is an earlier version of the UN's Expert Working Group's technical guidance and provides a 'conceptual framework for human and economic impacts for disaster loss accounting'. IRDR categorises human impact indicators into three different levels:

- Primary indicators: deaths, missing, injured, exposed, and economic loss as cumulative estimates without distinguishing between direct or indirect impacts.
- Secondary indicators: primary indicators can be further disaggregated into direct and indirect deaths or economic loss, which can be further refined into direct and indirect loss.
- Tertiary indicators: primary and secondary indicators can be further refined by accounting for gender, age, or the location where the injuries occurred, or direct economic losses can be broken down by sector (e.g., agricultural, residential). (IRDR, 2015, p. 7)

Table 3.2 Number of deaths classification

Number of deaths		
Direct deaths	Indirect deaths	
	Indirect immediate deaths	Indirect delayed deaths

Source Adapted by the Author from IRDR (2015)

The number of deaths can be subdivided to include secondary information such as direct or indirect deaths. Indirect deaths can be further subdivided into indirect immediate deaths and indirect delayed deaths (see Table 3.2).

The number of deaths is the sum of and indirect immediate deaths. The number of delayed indirect deaths is generally excluded (IRDR, 2015), which is also emphasised by the Expert Working Group's technical guidance (Paul, 2021; UNISDR, 2017). The number of deaths does not include missing persons (IRDR, 2015).

Direct deaths are defined as "persons who died as a direct result of a disaster" or during a disaster (e.g., crushed by a building during an earthquake, or drowned in a storm surge) (IRDR, 2015, p. 9).

Indirect deaths per se are not defined by IRDR (2015). Instead, they are defined as 'indirect immediate deaths' (see below). As such, some academic literature is invoked here to enlighten the readers. Indirect deaths are caused by "unsafe or unhealthy conditions that occur because of the disaster. These conditions include the loss or disruption of essential services, personal loss, and disruption of an individual's lifestyle" (Combs et al., 1999, p. 1125). 'Unsafe or unhealthy conditions' can be exacerbated by pre-existing health conditions and socio-economic factors which influence health—often referred to as social determinants of health (SDH[3]) (WHO, 2023). Socio-economic factors can increase the risk of contracting an infectious and non-communicable disease and associated morbidity and mortality by influencing access to care, care-seeking behaviours, and the overall quality of treatment received. The manner of indirect deaths can be natural, non-natural, or of other causes (Combs et al., 1999). Chiu et al. (2013, quoted in Paul, 2021, p. 99) similarly define indirect death to Combs et al. (1999): "if it was caused by unsafe or unhealthy conditions generated by the disaster (e.g. hazardous roads) or a loss or disruption of usual services (e.g. a power outage)".

Indirect deaths are of two types: indirect immediate and indirect delayed deaths.

Indirect immediate deaths "include persons who died of other causes (within days, weeks to months depending on the peril) that were the result of the disaster occurring such as traffic accidents during wildfire evacuations" (IRDR, 2015, p. 9).

Data sources used by global disaster loss databases often report direct deaths and indirect immediate deaths, but this is highly variable among the databases (IRDR, 2015). However, IRDR's Guidelines and Expert Working Group's technical guidance do not specify the time for indirect immediate deaths specific to hazard classification (UNISDR, 2017). Furthermore, the Expert Working Group's technical guidance

[3] SDH is the 'conditions in which people are born, grow, live, work, age, including the health system' (WHO, 2023).

mentions: "For the purposes of monitoring and reporting deaths for Target A of the Sendai Framework, it is recommended to focus on the direct causes of death that are more feasible to attribute, collect and report" (UNISDR, 2017, p. 10). Direct and indirect deaths are different for each hazard. In the case of an earthquake or other disaster, for instance, if the death occurs 24 h later, it is usually considered an indirect immediate death (Paul, 2021).

Indirect, delayed deaths "are caused by longer-term effects of a disaster and are only visible and measurable well after the disaster happened (years to decades), such as radioactive exposures after a tsunami event. Data sources less often report data sources. Indirect delayed deaths are not included in the registered number of deaths" (IRDR, 2015, p. 9). This exclusion is emphasised by the Expert Working Group's technical guidance (UNISDR, 2017, p. 10).

This study focuses on indirect, immediate, and direct deaths (purely based on circumstances) from the impact of COVID-19 and the economic lockdown in India. However, they are coined here as 'avoidable disaster deaths' (discussed below) using the same definitions. This is because these are standard definitions for the sector—more on this in Chap. 5.

Death Audits

According to the Expert Working Group's technical guidance, "a robust methodology to measure mortality" is vital. Therefore, studies on disaster deaths aim to focus on reliable and accurate measurements to reduce the rate of disaster deaths. Typically, considerations around such measurement include the attribution of death from exposure to a hazard and the temporal relationship between the hazard event and death, as mentioned above (Goswami, 2022; UNISDR, 2017). The health sector relies on specific information about the 'cause of death' and 'risk of death' from exposure to natural hazard disasters to effectively plan rapid emergency responses that save lives and reduce premature mortality before, during and after a disaster (Goswami, 2022; UNISDR, 2017).

According to Ritchie and Roser (2020), it is essential to understand what is meant by the *cause of death* and *risk factor* association with a premature death [original emphasis by the authors]. In the epidemiological framework of the Global Burden of Disease study, each death has *one* specific cause "that initiated a series of events leading to death" (GBD 2016 Mortality Collaborators, 2017, quoted in Ritchie and Roser, 2020). Paul (2021) and Jonkman and Kelman (2005) (alluded to in the previous chapter) understand the 'cause of death' in the same vein in the context of disasters. 'Cause of death' refers to the medical cause of death (e.g. trauma, drowning, or being hit by moving debris) and a death certificate specifies the exact medical cause of each disaster-induced fatality (Paul, 2021).

However, a death certificate is not issued in many lower-middle and low-income countries and is overly delayed (Paul, 2021; Ray-Bennett, 2018). As such, disaster deaths are not recorded in time by the national government's disaster database or the emergency database on disaster loss and damage generated by the UNDRR's DisInventar or CRED's EM-DAT. They do not fully represent the actual loss of lives (Gupta et al., 2022; Paul, 2021). Furthermore, when death certificates are issued, the likely cause is not attributed to the direct impact of the disaster. There are many reasons for this, including inconsistent methodology and terminologies practised at national and local levels, a lack of understanding of indirect disaster deaths, and challenges related to receiving the certificate of cause of death from a disaster (Gupta et al., 2022; Paul, 2021; Ray-Bennett, 2018). But the most important one is ex-gratia compensation that a deceased family member is entitled to receive from the state or national government in India (Ray-Bennett, 2017, 2018). Therefore, missing data or misattribution of disaster deaths, disaster-related deaths or indirect disaster deaths is not uncommon.

In India, the 'cause of death' indicates the underlying reason for the demise of an individual, ascertained by the Medical Certification of Causes of Death (MCCD), as per the 'Registration of Births and Deaths Act 1969' (Sections 10 (2 and 3)) (ORGI, 2019). The scheme of the 'Report on Medical Certification of Cause of Death' was introduced under the provisions of the Registration of Births and Deaths Act 1969. Through this mechanism, the Office of Registrar General and Census Commissioner of India (ORGI) obtains data on medically certified deaths. Data are collected, compiled and tabulated by the Offices of the Chief Registrars of Births and Deaths of the various States and Union Territories (ORGI, 2019). The attending physician fills out the Medical Certification of Causes of Death, and in case of medico-legal cases, the cause of death section is filled by the medical office, which performs a post-mortem based on the evidence noticed by him/her (Gupta et al., 2022). In case of disaster deaths, the deceased family members can receive an ex-gratia payment by the submission of two important additional documents[4]: one is the 'certificate of cause of death from natural calamity' and the second is 'panchnama' prepared by the local Revenue Official who provides a complete set of documentation that includes a post-mortem report (Gupta et al., 2022). These two documents jointly act as a source of information for identifying the cause of death—immediate, antecedent, and underlying cause (such as disaster) (Gupta et al., 2022).

The Medical Certification of Causes of Death aims to provide a reliable information base to generate mortality statistics in India, including deaths from natural hazards (ORG, 2009; Gupta et al., 2022). The latest Medical Certification of Causes of Death report for 2018 (ORG, 2020a) is based on medical registration, accounting for only 21.1% of registered deaths in India, indicating that 78.9% of registered deaths are not medically certified. The 'level of registration' of deaths in India (which varies between different States) in 2018 was 89.3% (ORG, 2020b). Thus, medically certified deaths represent a smaller percentage than 22% of the total deaths occurring in the country. This indicates that we do not know the cause of death for many deaths and uptake of this practice (certificate of cause of death and Panchnama) is low amongst the community/beneficiaries or by the deceased family members in India. This also

[4] This rule is prescribed under the State Disaster Response Fund and the National State Disaster Response Fund.

indicates that the actual number of disaster deaths is grossly under-represented in India (Gupta et al., 2022). This merits the questions: What is the way forward? How can we best capture or record the causes of death in India and other countries to reflect the actual number of lives lost from a disaster?

One way to do this is through the verbal autopsy (VA) tool promoted by the World Health Organization (WHO). A verbal autopsy involves interviewing living family members or other close associates of the decedent (Gomes et al., 2017). The verbal autopsies are structured interviews conducted by non-medical field workers and then analysed by physicians to ascertain the cause of death. Autopsies generally include medical history, key symptoms, and a half-page local language narrative of the family story of the symptoms and events leading to deaths (Gomes et al., 2017).

In India, the verbal autopsy tool has been used by the Registrar General, Ministry of Home Affairs for the project 'Million Death Study' (MDS) to study maternal and child death (Edwards & Haleem, 2014; Gomes et al., 2017). When deaths occur outside hospital premises or in facilities with limited diagnostic capability, a verbal autopsy is used to measure the cause of death patterns. Under this system, those close to the deceased (identified under the Sample Registration System) are interviewed and instructed to fill up a pre-designed form used by trained physicians to arrive at the cause of death. This system has been necessitated by the fact that most deaths in India, especially those in rural areas, are not medically certified. The platform enables physicians to review the verbal autopsy tool, ascertain the cause of death and categorise it within the standard international codes (ICDs) (AIIHPH, 2019). However, to the best of my knowledge, there are currently no verbal autopsy tools to study [in]direct disaster deaths.[5]

Another way to understand the cause of death from disaster is through the lens of 'circumstances' surrounding the deaths. These studies are led by researchers/academics and some of the indicators for circumstances they introduced were discussed in Chap. 2. These indicators are presented in Table 3.3 below (Jonkman & Kelman, 2005; Paul, 2021; Ray-Bennett et al., 2024). In the context of this study, this Matrix was used to capture the causes and circumstances of avoidable disaster deaths in the first phase of COVID-19 and the economic lockdown in India. The specific challenges of reporting medical causes of death (among others) for this study are discussed in the subsequent chapters.

Having discussed the concept and practice of disaster deaths and practices to ascertain circumstances and cause of death, the following section introduces the concept of 'avoidable disaster deaths' coined for this study.

[5] Between 2019 and 2021, I attended several meetings with the Indian Council of Medical Research, Odisha, and the Indian Institute of Public Health, Bhubaneswar, to plan a verbal autopsy tool for indirect disaster deaths. During these meetings, a review was conducted to identify the existing verbal autopsy tools in India. However, due to funding constraints, this plan could not be implemented.

Table 3.3 Matrix of cause and circumstances of death

Mechanism (direct vs. indirect deaths)	Location of deaths (indoor vs. outdoor)	Location of deaths (home or hospital or enroute)
Activity at the time of death	The timing of death (day vs. night)	Whether the deceased person was aware of the disaster (lack of judgement)
Whether s/he attempted to take safety measures	Gender of the deceased (male/female/other)	Age (0–19 years, 20–51 years, and > 51 years)
Medical cause of death		

Source Adapted by the Author from Paul (2021), Jonkman and Kelman (2005), Ray-Bennett et al. (2024)

Avoidable Disaster Deaths

The concept of 'avoidable disaster deaths' is distinctly different from 'disaster deaths' (see Table 3.4), yet they are complementary. This is because the Expert Working Group's technical guidance and guidelines on the concept of disaster deaths identify some pertinent terminologies and definitions, such as direct deaths and indirect deaths, which are relevant to avoidable disaster deaths (discussed above). Cause of death, circumstances of death, accountability of actors in reducing disaster deaths, and a timeline of deaths (discussed above) apply to the concept of avoidable disaster deaths, too. However, a few new components are introduced for this study to advance the theory and practice related to Sendai Framework's disaster deaths and disaster risk management practice. These components are:

- Disaster deaths may be classified as avoidable and unavoidable;
- Avoidable disaster deaths are sub-classified as amenable deaths, preventable deaths and disaster risk governance-related deaths;
- Avoidable disaster deaths include direct and indirect immediate deaths within 24 h of the event.

These novel components are delved further in this section. According to this approach, disaster deaths are either avoidable or unavoidable. Avoidable disaster deaths are amenable, preventable, and governance-related deaths (see Table 3.4). Unavoidable disaster deaths are those that occur every day due to natural and unnatural causes or due to underlying health conditions. These underlying causes can be exacerbated during a disaster. I focus first on the three types of avoidable disaster deaths. Amenable and preventable deaths are caused due a lack of timely and preventative measures, respectively. Governance-related deaths are caused due to weak

Table 3.4 Types of avoidable disaster deaths

Avoidable disaster deaths		
Amenable death	Preventable death	Governance-related death

Source Author

disaster risk governance measures. Therefore, tailored measures are required to reduce amenable, preventable, and governance-related avoidable deaths, which are discussed below.

Amenable Measures

Timely and effective healthcare interventions, including secondary prevention and treatment (i.e. after the onset of disease to reduce case fatality), are vital to reduce amenable or treatable avoidable deaths (Cooke, 2019; Price, 2021). Timeliness involves reducing waits and sometimes harmful delays for those who receive and give care (NAS, 2001). A lack of timeliness signals a lack of attention to and respect for the patient, which is not tolerated in consumer-centred systems in other service industries. It suggests that care has not been designed with the patient's welfare at the centre. In addition to emotional distress, physical harm may result, for example, from a delay in diagnosis or treatment that results in preventable complications. Effectiveness refers to providing services based on scientific knowledge to all who could benefit and refraining from providing services to those not likely to benefit (avoiding underuse and overuse) (NAS, 2001). Effectiveness also refers to care based on systematically acquired evidence to determine whether an intervention, such as a preventive service, diagnostic test, or therapy produces better outcomes than alternatives—including the alternative of doing nothing. Evidence-based practice requires that those who give care consistently avoid the underuse of effective care and the overuse of ineffective care, which is more likely to harm than help the patient (Chassin, 1997).

Unlike other health challenges dealing with few individuals deciding when, where and how, to seek care at health facilities, timeliness in the case of a pandemic such as COVID-19 depended largely on early case detection, while effectiveness involved instituting precise and prompt treatment. The importance of timely and effective health care cannot be overemphasised. Some of the indicators to better understand the importance of timeliness and effective health care in preventing COVID-19 deaths include (although not limited to): (i) the time interval between the onset of symptoms and seeking medical interventions (maximum of five days[6]) (Rothan & Siddappa, 2020); (ii) the time interval between the arrival of the patient and commencing treatment (minimum of 30 min to one hour)[7] (Baker & Finkn, 2020); (iii) active treatment of cases (application of COVID-19 treatment protocol[8]) without wasting time on

[6] A study in the first phase found that the symptoms of COVID-19 appear after an incubation period of approximately 5.2 days (Rothan & Siddappa, 2020).

[7] Delays in commencing treatment have caused deaths because patients present late, and in addition, factors such as delays occasioned by lack of bed space, those who do not have chances of survival such as severe congestive cardiac failure, etc. (Baker & Fink, 2020).

[8] Missed diagnosis in the case of unscreened patients will help accelerate the death of COVID-19 patients.

wrong diagnosis before correct treatment; and (iv) the higher the case detection, the earlier treatment needs to commence.

Preventable Measures

Public health interventions are vital to avoiding preventable deaths (i.e. before the onset of diseases or injuries to reduce incidence) (Cooke, 2019; Price, 2021). Health intervention "is an act performed for, with or on behalf of a person or population whose purpose is to assess, improve, maintain, promote, or modify health, functioning or health conditions" (WHO, 2020a). An intervention involves: Target (the entity on which the Action is carried out), Action (a deed done by an actor to a Target) and Means (the processes and methods by which the Action is carried out) (KSU, 2018). Public health interventions can be epidemiology and surveillance, outreach, screening, health teaching, social marketing, and policy development—among other things (KSU, 2018).

In light of COVID-19, WHO rolled out preventable interventions to help curb the spread of the disease (see Table 3.5). The goal of all countries was to control the pandemic by slowing down the transmission and reducing mortality associated with COVID-19 (WHO, 2020b). The global strategic objectives set in the first phase of COVID-19 are as follows:

(i) Mobilise all sectors and communities to ensure that every sector of government and society takes ownership of and participates in the response and preventing cases through hand hygiene, respiratory etiquette, and individual-level physical distancing.
(ii) Control sporadic cases and clusters and prevent community transmission by rapidly finding and isolating all cases, providing appropriate care, and tracing, quarantining, and supporting all contacts.
(iii) Suppress community transmission through context-appropriate infection prevention and control measures, population-level physical distancing measures, and proportionate restrictions on non-essential domestic and international travel.
(iv) Reduce mortality by providing appropriate clinical care for those affected by COVID-19, ensuring the continuity of essential health and social services, and protecting frontline workers and vulnerable populations.
(v) Develop safe and effective vaccines and therapeutics that can be delivered at scale and accessible based on need (WHO, 2020b).

A lot depends on adhering strictly to the interventions instituted and timely and effective health care to curb deaths from COVID-19. In other words, the number of active cases and deaths recorded reflects the effectiveness of strategies/interventions instituted, timeliness and effectiveness of health care and interventions, and so indicators such as confirmed cases, active cases, new cases, and new deaths are used to monitor these factors.

Table 3.5 Preventable strategy/interventions: the first phase

Strategy/ Intervention	Meaning	Rationale
Social distancing/ Physical distancing	Keeping space between yourself and other people outside of your home. To practice social or physical distancing: • Stay at least 6 feet (2 m) from other people (CDC, 2020) • Do not gather in groups • Stay out of crowded places and avoid mass gatherings	To reduce the chances of contacting COVID-19 through contacts, coughing, sneezing, droplets (breaking the chain of transmission)
Quarantine	To keep someone who *might* have been exposed to COVID-19 away from others	To reduce the chances of transmission
Isolate	To separate sick people from healthy people	
Hand hygiene	Regularly and thoroughly clean your hands with an alcohol-based hand rub or wash them with soap and water for at least 20 s	The activity kills viruses that may be on your hands
Respiratory hygiene	When coughing or sneezing, cover your mouth and nose with a bent elbow or tissue. Then, dispose of the used tissue immediately	The activity reduces/prevents the release of droplets into the air and the possibility of someone breathing in droplets, including the COVID-19 virus if the person coughing has the disease
Wearing of mask	Covering the nose and mouth with a mask	This activity prevents dissemination of virus by positive individuals through nasal and buccal droplets
Contact tracing	Process of monitoring persons who have been exposed to another person infected with a specific disease	To aid in early detection and prompt isolation of new cases
Lockdown	Restricting movement of people either locally or internationally	To reduce chances of transmission as the virus moves from person to person
Mass screening (active case search)	Reaching out to the community and systematically screening the population to find cases of COVID-19	This activity helps to reduce disease transmission by shortening patients' infectious periods, and earlier diagnosis and treatment improve patients' treatment outcomes
National emergency	A situation in which a government is empowered to perform actions or impose policies that it would normally not be permitted to undertake e.g. instituting a lockdown	To curtail the spread of the virus

Source Adapted by the author from CDC (2020) and WHO (2020c)

Disaster Risk Governance Measures

It is argued that effective disaster risk governance is essential to reduce avoidable amenable and preventable deaths. 'Government' and 'governance' are different terminologies (Alam & Ray-Bennett, 2021). The former embodies a hierarchical top-down arrangement, and the latter embodies a non-hierarchical arrangement incorporating the state and non-state actors who bring forth collective binding policies without superior authority (Alam & Ray-Bennett, 2021). Governance implies a reduction in the role of the state or the government while concurrently recognising the importance of non-state actors in the governance process. In other words, governance may be defined as integrating inter-sector and inter-governmental agencies that discharge services to the citizens (Alam & Ray-Bennett, 2021; Kapucu, 2012). The concept of disaster risk governance advocated by the Hyogo Framework addresses a similar idea of mainstream governance (Alam & Ray-Bennett, 2021).

The Hyogo Framework promoted the concept of disaster risk governance through two of its Priorities for Action: "Ensure that disaster risk reduction is a national and a local priority with a strong institutional basis for implementation" (Priority for Action 1), and "Strengthen disaster preparedness for effective response at all levels" (Priority for Action 5) (Alam & Ray-Bennett, 2021; UNISDR, 2005, p. 6). To promote effective disaster risk governance, the Hyogo Framework suggested a multi-sectoral approach and the devolution of disaster management activities, as well as the development of policies, strategies, knowledge and mutual learning through the Global Platform for Disaster Risk Reduction, the Regional Platforms for Disaster Risk Reduction, and regional cooperation (Alam & Ray-Bennett, 2021; Mizutori, 2020; UNISDR, 2005, 2015). In other words, disaster risk governance is understood as "the way in which public authorities, civil servants, media, private sector, and civil society at community, national and regional levels cooperate [collaborate, communicate and coordinate] in order to manage and reduce disaster and climate-related risks" ensuring "sufficient levels of capacity and resources are made available to prevent, prepare for, manage, and recover from disasters[9]" (Alam & Ray-Bennett, 2021; UNDP, 2012, p. 1). Disaster risk governance is currently continued through the Sendai Framework's Global Target E ("Substantially increase the number of countries with national and local disaster risk reduction strategies by 2020") (UN, 2015, p. 2) and the Priority for Action 2: "Strengthening disaster risk governance to manage disaster risk" (UN, 2015, p. 4). Target E mandated that every local government needs to have a disaster management plan in place by 2020. According to the UNDRR's Status Report 2020, 93 countries achieved Target E and 100 countries (52%) did not. As such, "to accelerate efforts to reach Target E by the end of 2020 remained an urgent [and yet an unmet] priority" for the UNDRR (UNDRR 2020, p. 7).

Ray-Bennett (2017, 2018) and Alam and Ray-Bennett (2021) found 20 practical components or characteristics generally used by researchers and practitioners to study disaster risk governance, summarised in Table 3.6.

[9] Since 2010, the United Nations Development Programme (UNDP) has been implementing disaster risk governance at the country level (UNDP, 2012; Alam and Ray-Bennett, 2021).

Table 3.6 Components of disaster risk governance

Characteristic	Descriptions
Accountability	Accountability means taking responsibility for implementing disaster risk governance principles
Participation	Participation of all stakeholders in different disaster risk management cycles
Collaboration	Collaboration means the application of networked coordination, collaboration and partnerships disaster risk management emphasising decentralised and flexible structures along with relevant administrative and service delivery adjustments
Transparency	Transparency implies that disaster risk management stakeholders are accountable for their actions to line management, interdepartmental organisations, and the community they serve. Accountability plays a significant role in the transparent delivery of services
Information sharing	Information is shared between different departments and communities during different disaster risk management cycles through formal and informal channels
Communication	Frequent and open lines of communication between different departments and communities may help disaster risk management. (Also see Chap. 2, the section on Complex Systems Approach, which focuses on communication failure)
Shared decision-making	Active engagement of horizontal-vertical agents and actors to make decisions for well-defined and mutually agreed goals and to implement such goals
Leadership	Strong leadership adds legitimacy and credibility to a collaborative process between organisations and the community
Shared resources	Sharing resources from different departments while working on disaster risk management may lead to the creation of something greater than any one individual or team could produce alone
Coordination	Coordination involves coordinating human personnel, the division of labour by function, stakeholder partnerships, and the coordination of the flow of 'core information'. Core information is early warning information and constitutes near-real-time information about an event's impact. (Also see Chap. 2, the section on Complex Systems Approach, which discusses on coordination failure)

Source Adapted by the Author from Ray-Bennett (2018), Alam and Ray-Bennett (2021)

However, not all these characteristics are feasible to study to assess the effectiveness of disaster risk governance during disasters. Furthermore, these characteristics are part and parcel of disaster management organisations (such as primary, secondary, and tertiary) undertaking various activities related to disaster risk governance. Tracking and analysing them all is not feasible (Alam & Ray-Bennett, 2021; Ray-Bennett, 2018). Therefore, selecting specific components suitable for the conceptual and institutional context is a way forward (Alam & Ray-Bennett, 2021; Ray-Bennett, 2018). As such, this study focussed on three components of disaster risk governance: coordination (C), communication (C), and collaboration (C). These components are not stand-alone. They interact and intersect in a complex

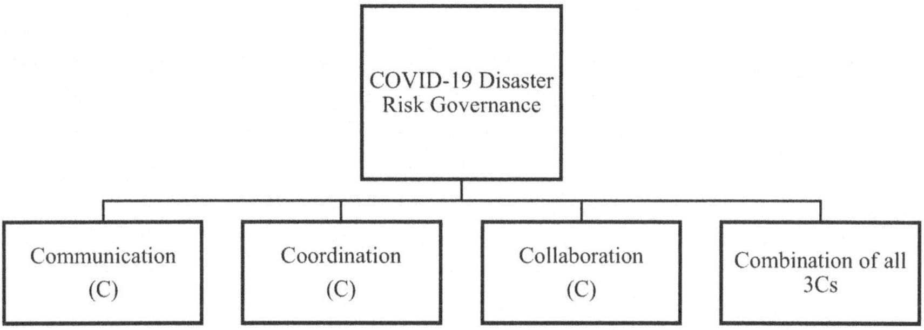

Fig. 3.1 Disaster risk governance for COVID-19. *Source* Author

way, as observed in this study (Chap. 5), and therefore, an additional component is introduced—coined here as 3Cs (please see Fig. 3.1). Therefore, coordination, communication, collaboration and 3Cs are collectively called COVID-19/disaster risk governance. Having discussed the components of avoidable disaster deaths, the ensuing section focuses on unavoidable disaster deaths.

Unavoidable Disaster Deaths

Some deaths are unavoidable during disasters. This is because disasters happen in an everyday context. Death is part of life. It is inevitable and occurs every day due to natural causes (old age), unnatural causes (poverty-related deaths, or other determinants of deaths (age, gender, ethnicity etc.)), injuries (snakebites, drowning, road traffic incidents etc.) or underlying health conditions (co-morbidities). These underlying causes can be exacerbated during or in the aftermath of a disaster because disasters create unhealthy and unsafe conditions, leading to indirect disaster deaths. Indirect disaster deaths can be immediate or delayed over a defined period of time—as discussed above.

Regardless of the disaster, a certain number of deaths occur each year in a particular region or country, and this is called the mortality rate. The annual mortality rate is defined as "a measure of the frequency of occurrence of death in a defined population during a specified interval" (CDC, 2012). These death rates are maintained by most countries, including India, and this is discussed in the next chapter. In the United Kingdom, the Office for National Statistics maintains death rates. These rates represent the country's annual hazards, i.e., the "proportion of people of each year of age, who do not reach their next birthday" (Spiegelhalter, 2020). These rates are maintained each year, even in the absence of a biological disaster such as the COVID-19 pandemic.[10] The annual mortality rate is distinct from excess mortality or excess deaths.

[10] According to some experts, COVID-19 approximately contributes a year's worth of risk. Here, risks refer to the average risk for persons of the relevant age, rather than the risks faced by the average person.

Excess deaths refer to a short-term rise in the mortality rate or the number of deaths occurring in a country or region because of events such as heat waves, pandemics, wars, and famines—among other hazards. Beaney et al. (2020, p. 330) define excess mortality as "an estimate of the additional number of deaths within a given time period in a geographical region (e.g., country), compared to the number of deaths expected (often estimated using the same time period in the preceding year or averaged over several preceding years)". These authors suggest excess deaths can be direct and indirect. This duality of excess deaths gained importance in the context of COVID-19 and the consequent economic lockdowns of 2020. During COVID-19, there was a growing focus on excess mortality due to the challenges it can overcome as a measure of comparison (Beaney et al., 2020).

The distribution of deaths (or unavoidable deaths in this context) is broken down into three broad categories by The *Global Burden of Disease*[11]—communicable diseases (CDs), non-communicable diseases (NCDs), and injuries. *Non-communicable diseases* cannot be passed from person to person[12]; *communicable or infectious diseases* are caused by a pathogen which can be passed from person to person; and *injuries* are a very wide category which includes accidents—such as car crashes—as well as intentional injuries like homicides, war deaths, and **suicides** [my emphasis] (Roser, 2021).

This study considers deaths from certain causes (non-communicable diseases, communicable diseases, and injuries) to be unavoidable. An example of this is malaria. Although malaria, along with other causes of unavoidable deaths, is undoubtedly avoidable in the long term, they are unavoidable in the short term, especially during the time of a disaster or when the disaster is unfolding (such as, during the economic lockdown period of COVID-19). They formed part of the context into which COVID-19 arrived and spread in India or another part of the world at the beginning of 2020. Thus, this study focuses only on avoidable direct and indirect deaths over ten months (short-term, 11 March–30 December 2020). All other causes of death are held constant as unavoidable for this study.

Conclusion

To conclude, Table 3.7 and Fig. 3.2 capture the components of avoidable disaster deaths discussed above. Subsequent chapters draw on the findings of the empirical data from India to make sense of the unavoidable and avoidable disaster deaths that took place in the first phase of COVID-19 and the economic lockdown.

---

[11] https://www.thelancet.com/gbd The Global Burden of Disease is a significant study on the 'causes of deaths and disease'.

[12] The two most common causes of death fall into this group: cancers kill 18% of people and cardiovascular diseases—such as stroke and ischemic heart disease—are responsible for one-in-three deaths in the world (Roser, 2021).

Table 3.7 Components of avoidable disaster deaths

Avoidable disaster deaths		
Avoidable		*Unavoidable (in the context of this study)*
Amenable death, preventable death, governance-related death		Communicable diseases, Non-communicable diseases, and injuries
Direct	Indirect	

Source Author

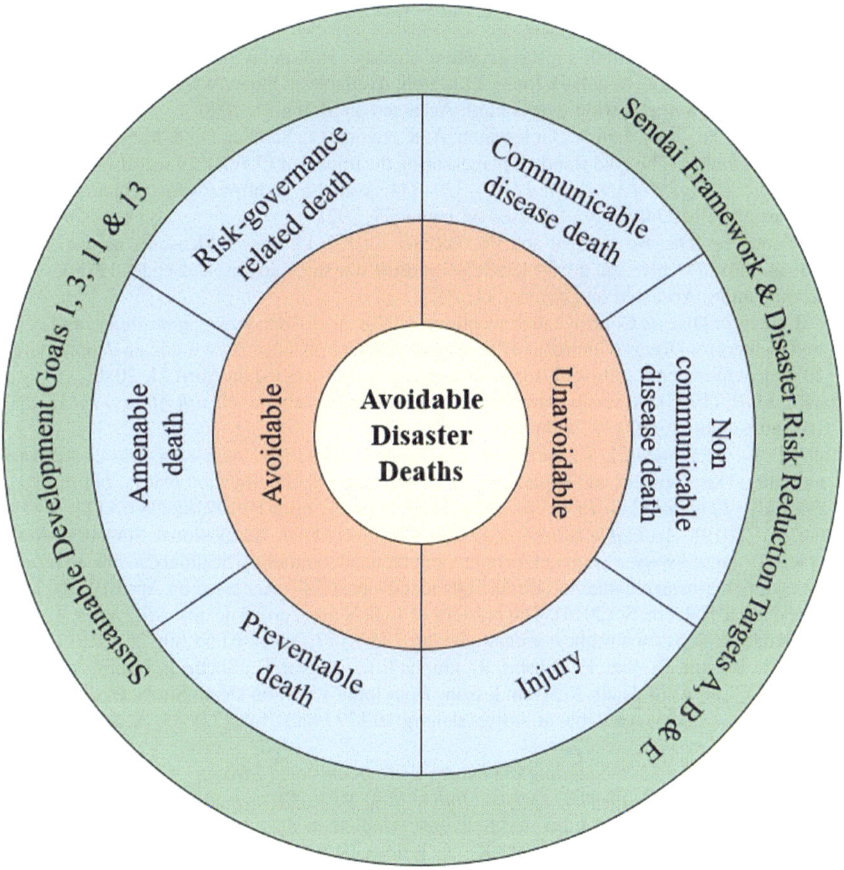

Fig. 3.2 Components of avoidable disaster deaths approach. *Source* Author

References

AIIHPH (All India Institute of Hygiene and Public Health). (2019). *MINERVA: Mortality in India established through verbal autopsy.* Available at: http://aiihph.gov.in/minerva-mortality-in-india-established-through-verbal-autopsy/

Alam, E., & Ray-Bennett, N. S. (2021). Disaster risk governance of local level administration in Rangamati, Bangladesh. *International Journal for Disaster Risk Reduction, 59*(2021), 1–13. https://doi.org/10.1016/j.ijdrr.2021.102220

Bahinipati, C. S., & Gupta, A. K. (2022). Methodological challenges in assessing loss and damage from climate-related extreme events and slow onset disasters: Evidence from India. *International Journal of Disaster Risk Reduction, 83*(2022), 103418. https://doi.org/10.1016/j.ijdrr.2022.103418

Baker, M., & Fink, S. (2020). It's just everywhere already': How delays in testing set back the U.S. coronavirus response. *New York Times*, 11 March. Available at: https://www.nytimes.com/2020/03/10/us/coronavirus-testing-delays.html. Accessed on March 11, 2020.

Beaney, T., Clarke, J. M., Jain, V., Golestaneh, A. K., Lyons, G., Salman, D., & Majeed, A. (2020). Excess mortality: The gold standard in measuring the impact of COVID-19 worldwide? *Journal of the Royal Society of Medicine, 113*(9), 329–334. Available at: https://www.ncbi.nlm.nih.gov/pmc/articles/PMC7488823/. Accessed on January 7, 2021.

CDC (Center for Disease Control and Prevention). (2012). Lesson 3: Measures of risk. Available at: https://archive.cdc.gov/#/details?url=https://www.cdc.gov/csels/dsepd/ss1978/lesson3/section3.html. Accessed on February 11, 2025.

CDC (Center for Disease Control and Prevention). (2020). *Social distancing, quarantine, and isolation.* Centers for Disease Control and Prevention. Available at: https://www.cdc.gov/coronavirus/2019-ncov/prevent-getting-sick/social-distancing.html. Accessed on April 21, 2020.

Chassin, M. R. (1997). Assessing strategics for quality improvement. *Health Affairs, 16*(3), 151–161. https://doi.org/10.1377/hlthaff.16.3.151

Combs, D. L., Quenemoen, L. E., Parrish, R. G., & Davis, J. H. (1999). Assessing disaster-attributed mortality: Development and application of a definition and classification matrix. *International Journal of Epidemiology, 1999*(28), 1124–1129. https://doi.org/10.1093/ije/28.6.1124

Cooke, A. (2019). *Avoidable mortality in the UK.* The Office for National Statistics. Available at: https://www.ons.gov.uk/peoplepopulationandcommunity/healthandsocialcare/causesofdeath/bulletins/avoidablemortalityinenglandandwales/2017. Accessed on April 19, 2022.

Edwards, C., & Haleem, S. (2014). The mystery of India's unrecorded deaths. *BBC News*, 12 July. Available at: https://www.bbc.com/news/health-28228177. Accessed on July 20, 2020.

Gomes, M., Begum, R., Sati, P., Dikshit, R., Gupta, P. C., Kumar, R., Sheth, J., Habib, A., & Jha, P. (2017). Causes of death: Relevant lessons from India's Million Death Study. *Health Affairs, 36*(11), 1887–1895. Available at: https://doi.org/10.1377/hlthaff.2017.0635. Accessed on July 19, 2020.

Goswami, N. (2022). *Understanding the causes and circumstances leading to snakebites during disasters in Ganjam district, Odisha, India.* MSc Risk, Crisis and Disaster Management Dissertation. University of Leicester, Leicester. [Unpublished].

Gupta, A. K., Chopde, S., Bindal, M. K., & Baidya, S. (2022). *Institutional mechanism for evaluating human deaths from disasters.* ResearchGate. Available at: https://www.researchgate.net/publication/362347714_Institutional_mechanism_for_evaluating_human_deaths_from_disasters. Accessed on April 4, 2023.

IRDR (Integrated Research on Disaster Risk). (2015). *Guidelines on measuring losses from disasters: Human and economic impact indicators* (IRDR DATA Publication No. 2). IRDR.

Jonkman, S. N., & Kelman, I. (2005). An analysis of causes and circumstances of flood disaster deaths. *Disasters, 29*(1), 75–97.

Kapucu, N. (2012). Disaster and emergency management systems in urban areas. *Cities, 29*, 541–549.

KSU (Kent State University). (2018). *A look at public health interventions*. Kent State University. Available at: https://onlinedegrees.kent.edu/college-of-public-health/community/public-health-interventions. Accessed on April 21, 2020.

Maini, R., Clarke, L., Blanchard, K., & Murray, V. (2017). The Sendai framework for disaster risk reduction and its indicators—Where does health fit in? *International Journal of Disaster Risk Science, 8*(2), 150–155.

Mizutori, M. (2020). Reflections on the Sendai framework for disaster risk reduction: Five years since its adoption. *International Journal of Disaster Risk Science, 11*, 147–151.

NAS (National Academies of Sciences). (2001). *Crossing the quality chasm: A new health system for the 21st century: Improving the 21st-century health care system*. National Library of Medicine. Available at https://www.ncbi.nlm.nih.gov/books/NBK222265/. Accessed on April 20, 2020.

ORG (Office of the Registrar General). (2009). *Report on causes of death in India 2001–2003*. Available at: http://www.cghr.org/wordpress/wp-content/uploads/Causes_of_death_2001-03.pdf. Accessed on August 9, 2020.

ORG (Office of the Registrar General). (2020a). *Report on medical certification of cause of death—2018*. Office of the Registrar General. Available at: http://crsorgi.gov.in/web/uploads/download/MCCD_Report-2018.pdf. Accessed on August 31, 2020.

ORG (Office of the Registrar General). (2020b). *Vital statistics based on the civil registration system 2018*. Available at: https://censusindia.gov.in/2011-Common/CRS_2018/crs2018_2007 2020.pdf. Accessed on August 9, 2020.

ORGI (Office of the Registrar General of India). (2019). *Report on the medical certification of cause of death*. Available at: http://censusindia.gov.in/2011-Documents/mccd_Report1/MCCD_Report-2017.pdf. Accessed on June 15, 2020.

Paul, B. K. (2021). *Disaster deaths: Trends, causes and determinants*. Routledge.

Price, M. (2021). Avoidable mortality in the UK: 2019. In *Statistical bulletin*. Office for National Statistics. Available at: https://www.ons.gov.uk/peoplepopulationandcommunity/healthand socialcare/causesofdeath/bulletins/avoidablemortalityinenglandandwales/2019. Accessed on September 27, 2023.

Ray-Bennett, N. S. (2017). Disasters, deaths and the Sendai goal one: Lessons from Odisha, India. *World Development, 103*(2018), 27–39. https://doi.org/10.1016/j.worlddev.2017.10.003

Ray-Bennett, N. S. (2018). *Avoidable deaths: A systems failure approach to disaster risk management*. Springer Nature.

Ray-Bennett, N. S., Clarke, K., & Mendez, D. (2022). Sendai framework's global Targets A and B: Opinions from the global platform for disaster risk reduction's ignite stage 2019. *International Journal for Disaster Risk Science, 13*(2022), 651–663. https://doi.org/10.1007/s13753-022-004 32-3

Ray-Bennett, N. S., Dissanayake, L., Ekezie, W., Macleod, L. (2024) Assessment of drowning data collection processes in low-and middle-income countries: A scoping review report. Avoidable Deaths Network and Royal National Lifeboat Institution.

Ray-Bennett, N. S., Mendez, D., Alam, E., & Morgner, C. (2020). Inter-agency collaboration in natural hazard management in developed countries. In B. Gerber (Ed.), *Oxford research encyclopedia of natural hazard science*. Oxford University Press. https://doi.org/10.1093/acrefore/9780199389407.013.176

Ritchie, H., & Roser, M. (2020). *Causes of death*. OurWorldinData.org. Available at: https://archive.vn/qHbfi#selection-15459.0-15459.39. Accessed on March 25, 2021.

Roser, M. (2021). *Causes of death globally: What do people die from?* OurWorldinData.org. Available at: https://ourworldindata.org/causes-of-death-treemap. Accessed on October 22, 2024.

Rothan, H., & Siddappa, H. (2020). The epidemiology and pathogenesis of coronavirus disease (COVID-19) outbreak. Available at https://www.researchgate.net/publication/339515532_The_epidemiology_and_pathogenesis_of_coronavirus_disease_COVID-19_outbreak. Accessed on April 20, 2020.

Sen, A. (2020). Listening as governance. *CPNN*, 21 June. Available at: https://cpnn-world.org/new/?p=20650. Accessed on June 21, 2020.
Spiegelhalter, D. (2020). 'How much 'normal' risk does Covid represent?' Winton Centre, University of Cambridge. Available at: https://medium.com/wintoncentre/how-much-normal-risk-does-covid-represent-4539118e1196. Accessed on 11 February, 2025.
UN (United Nations). (2015). *Sendai framework for disaster risk reduction 2015–2030*. UNDRR. Available at: http://www.unisdr.org/we/inform/publications/43291. Accessed October 9, 2024.
UN (United Nations) General Assembly. (2016). *Report of the open-ended intergovernmental expert working group on indicators and terminology relating to disaster risk reduction*. UN General Assembly in its resolution A/RES/71/276. Available at: https://www.unisdr.org/we/inform/publications/51748. Accessed on July 17, 2022.
UNDP (United Nations Development Programme). (2012). *Issue brief: Disaster risk governance*. UNDP. Available at: https://www.preventionweb.net/publications/view/29974. Accessed on March 16, 2022.
UNDRR (United Nations Office for Disaster Risk Reduction). (2020). *Status report on Target E implementation*. UNDRR. https://www.undrr.org/publication/status-report-target-e-implementation-2020. Accessed on April 4, 2023.
UNISDR (United Nations Office for Disaster Risk Reduction). (2005). Hyogo framework for action 2005–2015: Building the resilience of nations and communities to disasters. In *Proceedings of World Conference on Disaster Reduction, 18–22 January 2005, Kobe, Japan*. http://www.unisdr.org/we/inform/publications/1037. Accessed on July 17, 2022.
UNISDR (United Nations Office for Disaster Risk Reduction). (2017). *Technical guidance for monitoring and reporting on progress in achieving the global targets of the Sendai Framework for Disaster Risk Reduction: Collection of technical notes on data and methodology*. UNISDR. https://www.unisdr.org/we/inform/publications/54970. Accessed on July 17, 2022.
WHO (World Health Organization). (2020a). *COVID-19 strategy update*. Available at: https://www.who.int/publications/m/item/covid-19-strategy-update. Accessed on April 21, 2020.
WHO (World Health Organization). (2020b). *International Classification of Health Interventions (ICHI)*. Available at: https://www.who.int/classifications/ichi/en/. Accessed on April 21, 2020.
WHO (World Health Organization). (2020c). *WHO director-general's opening remarks at the media briefing on COVID-19—11 March 2020*. Available at: https://www.who.int/director-general/speeches/detail/who-director-general-s-opening-remarks-at-the-media-briefing-on-covid-19---11-march-2020. Accessed on April 20, 2020.
WHO (World Health Organization). (2023). *Social determinants of health*. Available at: https://www.who.int/health-topics/social-determinants-of-health#tab=tab_1. Accessed on September 27, 2023.

Chapter 4
Unavoidable Deaths in India

> *Reliable statistics describe the reality of people's everyday lives. This picture of society tells us where the poor are, why they are poor and what their lives are like. [...] It highlights where resources are most needed, and provides the means to track progress and assess the impact of different policies.*
> —PARIS21 and OECD/DCD (2013)

Abstract This chapter presents the findings of the empirical data on the trends of unavoidable deaths in India before the COVID-19 pandemic. The timeline for this data review is from 2007 to 2018. The timeline was selected based on data availability and in consultation with the Vital Statistics Division. The sources of unavoidable deaths include four reports published by the Government of India's Office of the Registrar General and Census Commissioner of India (ORGI) and two reports by the National Crime Records Bureau (NCRB). These data are vital statistics for India and provide up-to-date information on unavoidable deaths. By doing so, this study also uses the broad categories of cause of death, categories of fatalities and categories of crimes advocated by ORGI and NCRB in these reports. These broad categories then act as a reference point for this study to understand and capture avoidable disaster deaths from the impact of COVID-19 and the economic lockdown in India in the first phase.

Keywords Accidental deaths and suicides · Avoidable disaster deaths · Cause of death · Unavoidable deaths · Sample registration system · India

Introduction

Deaths are both unavoidable and avoidable, as discussed in Chap. 3. This chapter presents the findings of the empirical data on the trends of unavoidable deaths in India, especially before the COVID-19 pandemic. In simple words, unavoidable deaths are understood as deaths that occur annually. In India, the Government of India's Office

N. S. Ray-Bennett, *Avoiding Disaster Deaths*,
https://doi.org/10.1007/978-3-031-81300-9_4

Table 4.1 Reports reviewed for this study

Reports	Published by
Medical Certification of Cause of Death 2017	Office of the Registrar General and Census Commissioner of India[2] (2019)
Causes of Death Statistics 2007–2013 and 2010–2013	Office of the Registrar General and Census Commissioner of India (2014b, 2014c)
Sample Registration System Statistical Report 2018	Office of the Registrar General and Census Commissioner of India (2020)
Accidental Deaths and Suicides in India 2018	National Crime Records Bureau, Ministry of Home Affairs (2019)
Crime in India 2018	National Crime Records Bureau, Ministry of Home Affairs (2019)

of the Registrar General and Census Commissioner of India (ORGI) and the National Crime Records Bureau (NCRB) keep the statistics of unavoidable deaths.

At the outset, it is important to emphasise that many unavoidable deaths in India are avoidable (discussed in Chap. 3). However, for this study, they are considered unavoidable. This chapter provides an overview of unavoidable deaths between 2007 and 2018 using statistics, death classification, and causes and circumstances of deaths. This empirical data then acts as a reference point to explain how unavoidable death trends continued and capture them using the classification of deaths during the economic lockdown in India's first phase of COVID-19.

Statistics

Statistics on unavoidable deaths in India are limited and inconsistent. There are but a few sources upon which to rely. Six relevant sources were reviewed for this study (see Table 4.1), published by different departments under the Government of India's Ministry of Home Affairs[1] (ORGI, 2021a). Of these six reports, four are published by the Office of the Registrar General and Census Commissioner of India and two by the National Crime Records Bureau.

Office of the Registrar General and Census Commissioner of India (ORGI) prepares and publishes the causes of death reports. In addition, it conducts the Housing and Population Census in India under the Census Act 1948 and the Rules; coordinates the functioning of the Civil Registration System under the Registration of Births and Deaths Act, 1969; implements the Sample Registration System on a half-yearly basis; prepares the National Population Register under the provision

[1] The Ministry of Home Affairs (MHA) discharges multifarious responsibilities, the important among them being—internal security, border management, Centre-State relations, administration of Union Territories, management of Central Armed Police Forces, disaster management, etc." (ORGI, 2021a).

[2] https://censusindia.gov.in/nada/index.php/catalog/40070.

contained in Citizenship Rules 2003 framed under the ambit of the Citizenship Act 1955; and conducts the Mother Tongue Survey for languages consistently returned across two and more census decades (ORGI, 2021b).

National Crime Records Bureau compiles and publishes the national crime statistics i.e., crime in India, accidental deaths and suicides and prison statistics. It also assists in developing the capacity of the states. Established in 1986, the National Crime Records Bureau is based in New Delhi (Director's Desk, 2022) and functions as a repository of information on crime and criminals based on the recommendations of the Tandon Committee, National Police Commission (1977–1981), and the Ministry of Home Affairs' Taskforce (1985). National Crime Records Bureau monitors, coordinates and implements the Crime and Criminal Tracking Network and Systems and manages the National Digital Police Portal, National Database of Sexual Offender Counterfeit Currency Information and Management System, Integrated Monitoring on Terrorism, Crime-Multi Agency Centre, and the Center Finger Print Bureau. The *Accidental Deaths and Suicides in India* (NCRB, 2019a) and *Crime in India* (NCRB, 2019b) reports are the two relevant reports selected for this study. They are reviewed following the Office of the Registrar General's reports.

These and older editions of the selective reports are available on the websites of the Office of the Registrar General and Census Commissioner of India and the National Crime Records Bureau. Only the latest reports were selected for review. Also, these reports were selected in consultation with the Senior Research Officer of the Vital Statistics Division and the Deputy Registrar General of India in 2020. The statistical findings in these reports vary between Indian States and within States (between rural and urban areas). As such, these data have some limitations and they have been mentioned wherever this is applicable. As mentioned in the previous chapter, not all deaths are registered; thus, even the total registered deaths do not cover all the annual mortalities in India.

According to a study conducted by the World Health Organization (WHO), the quality of death registration in India was ranked "very low" for the period between 2005 and 2015. The scale categories were high, medium, low, and very low (WHO, 2017, p. 84). Regarding 'usability', which is "defined as the percentage of all deaths, which are registered with meaningful cause-of-death information" India scored circa 10% (WHO, 2017, p. 7). Despite the limitations of data as well as 'very low' quality of data according to WHO's evaluation, these are the data available for reference. They are vital statistics for India and for this study. Furthermore, the causes, circumstances, and classifications of deaths are vital statistics to understand the annual mortality rate, excess deaths vis-à-vis the impact of COVID-19 and how some unavoidable deaths extended and manifested during the lockdown. Although these statistics are numbers and lack a person's face and circumstances surrounding death, the subsequent chapters aim to overcome this by providing real cases of avoidable disaster deaths. In doing so, this chapter and the subsequent ones are complementary in making sense of some of the circumstances under which avoidable disaster deaths occurred from the impact of COVID-19 and the economic lockdown in India.

Medical Certification of Cause of Death

To capture reliable cause-specific mortality statistics, the scheme of Medical Certification of Cause of Death (MCCD) was introduced in the country under the provisions of the Registration of Births and Deaths (RBD) Act, 1969 (ORGI, 2019). Since then, the Medical Certification of Cause of Death has been operational in the country. Under this scheme, the Office of the Registrar General obtains data on medically certified deaths as collected, compiled and tabulated by the Offices of the Chief Registrars of Births and Deaths of the States/UTs (ORGI, 2019).

Data for the annual Medical Certification of Cause of Death is collected in the prescribed forms designed by the WHO—form 4 for hospital deaths and 4A for institutional deaths. These are completed by medical professionals who attended to the deceased patient at the time of treatment, who then send these to the concerned Registrars of Birth and Deaths to be transmitted to the Chief Registrar's Office of the State. The data is tabulated based on the Tenth Revision of the International Classification of Disease-10, after which it is sent to the Office of the Registrar General and Census Commissioner of India for consolidation at the national level (ORGI, 2019).

A significant limitation of the Medical Certification of Cause of Death report is that the data on causes of death, on which it is based, is mainly limited to deaths occurring in hospitals and other medical institutions, mostly in urban areas (ORGI, 2019). The overall coverage is low, as approximately 20% of total registered deaths are medically certified despite the extension of coverage. In 2014, the Government of India extended the scheme's coverage to all medical institutions, including government, private, and non-profit medical institutions, to bridge the data gap on India's cause-specific mortality profile. To support this initiative, the Registrar General of India also issued a circular to all Chief Registrars of Births and Deaths in the country (ORGI, 2014a). Consequently, there has been significant long-term growth in medically certified cases over the last three decades, from 14.7% in 1991 to 22% in 2017 (ORGI, 2019). There has also been a consistent but slow increase in the number of Indian States implementing the Medical Certification of Cause of Death scheme. However, the degree of efficiency of medical certification between States varies. In Goa, 100% of the registered deaths were medically certified in 2017, while the lowest percentage was seen in Jharkhand at 4.6% (ORGI, 2019).

During data collection in 2020, the ***Medical Certification of Cause of Death 2017*** was the latest report (ORGI, 2019).[3] According to this report, the eight leading cause-groups of death constituted 87% of the total medically certified deaths, and these are:

1. Diseases of the circulatory system (34%)
   (i) Diseases of pulmonary circulation and other forms of heart diseases (12%)
   (ii) Ischemic heart diseases (10.9%)

[3] Corroborated by the Senior Research Officer of the Vital Statistics Division of ORGI (8 June 2020).

2. Certain Infectious and Parasitic diseases (10.4%)
   (i) Septicaemia (4.6%)
   (ii) Tuberculosis (2.9%)
3. Diseases of the respiratory system (9.2%)
   (i) Pneumonia (1.4%)
   (ii) Asthma (0.76%)
4. Neoplasm (cancer) (6.4%)
   (i) Malignant neoplasm of digestive organs (1.3%)
5. Conditions originating in the perinatal period (5.8%)
6. Injury, poisoning and other consequences of external causes (5.8%)
7. Endocrine, nutritional and metabolic diseases (5.3%)
8. Symptoms, signs and abnormal clinical and laboratory findings not elsewhere classified (10.1%) (ORGI, 2019)

"Diseases of the Respiratory System" caused 129,618 fatalities (9.2%) of the total medically certified deaths, of which influenza accounted for 2037 deaths (0.15%), and pneumonia accounted for 20,207 deaths (1.43%) of all medically certified deaths (ORGI, 2019). "Injury, poisoning and other consequences of external causes", although one of the eight leading causes of death, fell from 11.3% in 2000 to 5.8% in 2017. Other unspecified effects of external causes and certain early complications of trauma is a category that includes the International Classification of Disease -10 codes T33-T35 and T66-T73. This category accounted for 6227 deaths (0.44%) of medically certified deaths (ORGI, 2019). This is relevant to this study because starvation, T73.0, is included in this category. Unfortunately, a more detailed breakup of death per the International Classification of Disease-10 code is unavailable.

Causes of Death Statistics Using Verbal Autopsy

During data collection in 2020, the *Causes of Death Statistics 2007–2013* (ORGI, 2014b) and *Causes of Death Statistics 2010–2013* (ORGI, 2014c) were the latest reports[4] and they are reviewed in this section. These reports result from a special initiative undertaken by the Office of the Registrar General and Census Commissioner of India to provide reliable, representative and relevant data on deaths caused by diseases, injuries and risks at the national/state level (ORGI, n.d[5]). It also aims to

[4] Corroborated by the Senior Research Officer of the Vital Statistics Division of ORGI (8 June 2020). In another email exchange with the Deputy Registrar General of India, we were informed that a report for 2014-16 and year-on-year data is expected to be published and will be available by mid-2020 (25 June 2020).

[5] Cause of death in India 2010–2013 presentation, slide 3.

fill the gap left by the report on *Medical Certification of Cause of Death* (discussed above), which accounts for approximately 22% of the total registered deaths.

The causes of death statistics are based on the special survey of death findings under the Sample Registration System (SRS) using the verbal autopsy tool. The use of this tool has increased from 91% in 2001 to 98% in 2010 (ORGI, 2014c). Since the *Causes of Death* are based on the Sample Registration Survey, the *Sample Registration Survey Statistical Report* (ORGI, 2020a) and a few bulletins were also reviewed. The Sample Registration Survey is a "large-scale demographic survey for providing reliable annual estimates of infant mortality rate, birth rate, death rate and other fertility and mortality indicators at the national and sub-national levels" (ORGI, 2020b, p. 1). Covering a sample population of 8.1 million, it is one of the largest demographic surveys in the world. "Dual recording" is a unique feature of its system, whereby continuous enumeration and retrospective half-yearly surveys are conducted. This is followed by matching the two records to ensure correct data and subsequent field verifications of unmatched or partially matched cases. Thus, the system provides for a cross-check on the accuracy and completeness of the events of birth and death listed by the two independent functionaries (ORGI, 2020b). The details of the four selected reports are provided below.

Causes of Death Statistics 2007–2013: The causes of death are divided into four broad categories: communicable, maternal, perinatal, and nutritional conditions; non-communicable diseases; injuries; and symptoms, signs, and ill-defined conditions (ORGI, 2014b). These categories are broader than the Global Burden of Disease's three broad categories of communicable, non-communicable, and injuries introduced in Chap. 3. The report consists of five tables corresponding to the years 2007–2013 for India as well as bigger individual States and Union Territories (ORGI, 2014b):

(i) Distribution of Death by Major Cause Groups
(ii) Distribution of Death by Major Cause Groups in Urban and Rural Areas
(iii) Distribution of Death by Age and Sex
(iv) Distribution of Death by Age and Sex in Urban and Rural Areas
(v) Top 10 Causes of Death for all ages.

The bigger States and Union Territories that have been included in the report are Andhra Pradesh, Assam, Bihar, Chhattisgarh, Delhi, Gujarat, Haryana, Jammu and Kashmir, Jharkhand, Karnataka, Kerala, Madhya Pradesh, Maharashtra, Odisha, Punjab, Rajasthan, Tamil Nadu, Uttar Pradesh, Uttarakhand and West Bengal (ORGI, 2014b).

According to this report, between 2007 and 2013, 30.15% of the Indian population died of communicable, maternal, perinatal and nutritional conditions; 48.69% of the population died of non-communicable diseases; 10.54% perished from injuries; and 10.62% died of symptoms, signs and ill-defined conditions (ORGI, 2014b). These deaths collectively form the unavoidable deaths in the country according to the avoidable disaster deaths approach. The number of these deaths and the percentage varies by gender, age, and location (rural and urban)—for more details, please see ORGI (2014b). Furthermore, almost half of the deaths are due to non-communicable diseases (48.69%), and about 10.54% of deaths are from injuries, which are causes for concern. This has been the trend of unavoidable deaths for the period 2007–2013 and likely continued with some variations (see below) up until the spread of the COVID-19 pandemic in March 2020 and throughout the lockdown period. During the first wave of COVID-19, deaths from injuries were stark—and they are discussed in the next chapter.

Causes of Death Statistics 2010–2013: This report provides statistics on deaths under 14 headings, which are: (i) Overall Causes of Death Distribution (by gender); (ii) Distribution of Deaths by Major Cause Groups; (iii) Distribution of Deaths by age and gender; (iv) Top 10 Causes (all ages); (v) Top 10 Causes: below 29 days; (vi) Top 10 Causes: above 1 year; (vii) Top 10 Causes: 1–4 years; (viii) Top 10 Causes: 0–4 years; (ix) Top 10 Causes: 5–14 years; (x) Top 10 Causes: 15–29 years; (xi) Top 10 Causes: 30–69 years; (xii) Top 10 Causes: 70 + years; (xiii) Deaths from Special Conditions; (xiv) Top 10 Causes (all ages, major regions) (ORGI, 2014c).

The report covers six regions of India, including *North* (Chandigarh, Delhi, Haryana, Himachal Pradesh, Jammu & Kashmir, Punjab and Uttarakhand); *North East*: (Arunachal Pradesh, Assam, Manipur, Meghalaya, Mizoram, Nagaland, Sikkim and Tripura); *East*: (Bihar, Jharkhand, Orissa and West Bengal); *Central*: (Chhattisgarh, Madhya Pradesh, Rajasthan and Uttar Pradesh); *West*: (Dadra & Nagar Haveli, Daman & Diu, Goa, Gujarat and Maharashtra); *South*: (Andhra Pradesh, Andaman & Nicobar Islands, Karnataka, Kerala, Lakshadweep, Puducherry and Tamil Nadu) (ORGI, 2014c).

Since the category of 'injuries' is important to this study, about 10.7% of fatalities were reported between 2010 and 2013 (ORGI, 2014c). Of these, 'Motor Vehicle Accidents' formed 27.5%, fatalities caused by 'Other Than Motor Vehicle Accidents' formed 44%, Suicide accounted for 24.6%, reasons 'Other Than Suicide' formed 2.85% and Injuries of 'Undetermined Intent' accounted for 1% of fatalities. Respiratory illnesses (both communicable and non-communicable diseases) made up 11.56% of the fatalities during the same period (ORGI, 2014c). There is no specific information on starvation deaths, although according to the International Classification of Disease (ICD)-10 coding, Starvation is coded as T73.0. However, starvation may be categorised here as malnutrition deaths (ICD10 code E40-46).

Sample Registration Survey Statistical Report 2018: According to this report, the Indian death rate is 6.2, varying from 6.7 in rural areas to 5.1 in urban areas (ORGI, 2020b). Similarly, the death rate varies between and within different States. For instance, Odisha[6] is placed at 7.3 at the State level, 7.6 for rural areas and 5.9 for urban areas (ORGI, 2020b). This report also contains data on medical attention received before death. In 2018, 29.4% of deaths occurred at government hospitals, 18.5% occurred at private hospitals, and qualified professionals attended 33% of fatalities, while untrained functionaries and others attended 19.2% (ORGI, 2020b). These figures vary between States and within States between rural and urban areas.

[6] Now and then, Odisha is used as a comparative reference point. This is because the ADN India Hub is located there. Also, Odisha has been my research site since 1999.

Many more deaths are attended by untrained functionaries in rural than urban areas and occur at private hospitals more in urban than rural areas.

Accidental Deaths, Suicides and Crimes

After reviewing the relevant Office of the Registrar General and Census Commissioner of India reports, this section provides details of two more vital statistical reports published by the National Crime Records Bureau: Accidental Deaths, Suicides, and Crimes in India.

Accidental Deaths and Suicides in India is an annual report published by the National Crime Records Bureau since 1967 (NCRB, 2019a). The latest edition, which was published in 2019 during the data review stage, covers the year 2018. It provides in-depth statistical information on various categories of fatalities caused by accidents, suicides, and injuries in India. The National Crime Records Bureau compiles information from the various States and Union Territories in the prescribed format into this report. An accident is defined as "an unplanned event that results in injury, deaths, damage to property or some other losses" (NCRB, 2019a, p. 1). For statistical purposes, accidental death consists of two broad categories: (a) forces of nature and (b) other causes. Forces of nature include the following causes of death (NCRB, 2019a): (i) Avalanche; (ii) Exposure to Cold; (iii) Cyclone; (iv) Tornado; (v) Tsunami; (vi) Earthquake; (vii) **Epidemic** [my emphasis]; (viii) Flood; (ix) Heat/ Sun Stroke; (x) Landslide; (xi) Lightning; (xii) Torrential Rain; (xiii) Forest Fire; (xiv) Causes other than the above.

There was an increase of 18.2% in accidental deaths (from 6891 deaths to 8145 deaths) due to forces of nature and an increase of 2.0% in deaths (from 404,933 deaths to 412,959 deaths) due to other causes during 2019 as compared to 2018 (NCRB, 2019a). Of the above subcategories, the most relevant to this study is Epidemic, which caused a total of 3 deaths in 2017 and 0 deaths in 2018. Therefore, deaths that occurred from the impact of COVID-19 and the economic lockdown in the first phase are excess deaths and avoidable disaster deaths (see Chap. 5).

The "Other Causes" category includes deaths other than those caused by the forces of nature. It includes deliberate or negligent conduct on the part of human beings. The sub-categories within "Other Causes" of accidental deaths are as follows (NCRB, 2019a):

(i) Air Crash
(ii) Ships Accidents
(iii) Collapse of Structure
(iv) Drowning
(v) Electrocution
(vi) Accidental Explosion
(vii) Falls
(viii) Factory/Machine Accidents

(ix) Accidental Fire
(x) Firearm
(xi) Mines or Quarry Disaster
(xii) Traffic Accidents
(xiii) Stampede
(xiv) Sudden Deaths
(xv) Deaths of Women During Pregnancy
(xvi) Deaths Due to Consumption of Illicit/Spurious Liquor
(xvii) Killed/Injured by Animals
(xviii) Poisoning
(xix) Suffocation
(xx) Drug Overdose
(xxi) Other than Above Causes
(xxii) Causes Not Known.

In 2018, 404,933 persons died of "Other Causes" of accidental deaths, which is a 1.8% increase over 2017. Of the above subcategories, the most relevant to this project is Traffic Accidents (xii), which consists of three subcategories: (i) Road Accidents, (ii) Railway Crossing Accidents and (iii) Railway Accidents. Traffic Accidents account for 178,832 deaths (44.2%) of all deaths resulting from "Other Causes". January and May accounted for the country's maximum traffic accidents. In terms of time, the maximum number of accidents was reported between 6.00 pm and 9.00 pm (NCRB, 2019a).

A total of 24,545 deaths were caused by railway accidents in 2018, which is 6.06% of all deaths from "Other Causes". About 68.4% of railway accidents fell under the category "Fall from trains/collision with people on track". Railway Crossing Accidents caused 1507 deaths in 2018 (0.37%) of deaths from "Other Causes". Uttar Pradesh had the highest number of deaths, accounting for 656 cases (44.3%) of all Railway Crossing Accidents (NCRB, 2019a). The report also provides the causes of suicides. About 134,516 suicides were reported in India in 2018, showing an increase of 3.6% over the previous year (NCRB, 2019a).

***Crimes in India 2018*:** The annual report on Crimes has been published by the National Crime Records Bureau since 1953 (NCRB, 2019b). It contains comprehensive information on registered cases and their disposal, as well as arrested persons and their disposal. This is categorised under the various headings of the Indian Penal Codes and Special and Local Laws. This information is disaggregated by age and sex. It also contains information on custodial deaths, complaints against the police and injuries and casualties of police and civilians. The 2018 edition, published in 2019, includes three volumes with dedicated chapters on various noteworthy subcategories (NCRB, 2019b).

Complaints against police personnel, violence at the hands of the police and custodial deaths are not occurrences unique to the year 2020 during COVID-19 or to the circumstances caused by the pandemic, as evidenced by dedicated subsections in the report. The National Crime Records Bureau, however, declares a few limitations of this data, and they are as follows:

1. Since this report caters to the "Principal Offence Rule", the count under each crime heading may be under-reported.
2. The Bureau does not capture the reasons behind crimes committed; only those cases recorded by the police are included in the report.
3. The crime rates of States and metropolitan cities are not comparable.
4. The Crime Rate is not calculated for city-based crimes under 'Children, Scheduled Castes/Scheduled Tribes and Senior Citizen' due to the unavailability of city population figures.
5. Data clarifications were pending from the states of West Bengal, Assam, Arunachal Pradesh, Meghalaya, Sikkim, and the city of Kolkata, and they may be treated as provisional. (NCRB, 2019b, p. v)

Deaths in police custody for the year 2018 were divided into two categories: (a) the category of "persons not on remand" (which includes those arrested and yet to be produced in court) saw a total of 48 reported deaths in India; and (b) the category of "persons in remand" (which includes those in police/judicial remand) saw 24 deaths in India. The causes of such deaths include (i) suicide, (ii) deaths due to illness/ death in hospitals during treatment, (iii) injuries sustained before police custody, (iv) injuries sustained due to physical assault by police, (v) while escaping from custody, (vi) road accidents/journey connected with the investigation and lastly, (vii) others (NCRB, 2019b). In 2018, 14 civilians were killed in police firing, and 30 were injured; four civilians were killed in a police *lathi-charge* (*aka* baton charge), and 245 were injured. The highest frequency of injuries from *lathi-charge* was 125 in Maharashtra (NCRB, 2019b).

Questions of medical negligence appeared in the context of COVID-19. Medical negligence is also not a concern specific to the circumstances caused by COVID-19. The *Crime in India 2018* report sheds some light on this category. In 2018, about 218 cases of medical negligence were reported and 363 total cases for investigation (including pending cases from the previous year, cases reported in 2018 and reopened for investigation) (NCRB, 2019b). In the same period, 143 persons were arrested for medical negligence, where 137 were male and six were female. However, the limitation of this data—as mentioned above, they do not represent the whole of India, as clarifications were pending from West Bengal, Assam, Arunachal Pradesh, Meghalaya, and Sikkim (NCRB, 2019b).

Conclusion

To conclude, this chapter reviewed four reports published by the Office of the Registrar General and Census Commissioner of India and two reports published by the National Crime Records Bureau to demonstrate the trends of unavoidable deaths that occur in India annually or over a range of years in the pre-COVID-19 phase. Although these deaths are coined as unavoidable deaths for the convenience of this study, they are avoidable. For instance, many high-income and low-middle-income countries have avoided smallpox (communicable disease) deaths (Whitty, 2017). Likewise, deaths from non-communicable diseases and injuries are avoidable over time.

The *Causes of Death Statistics 2007–2013* report provides four broad categories of 'cause of death', of which 'nutritional deficiencies' under the category of 'communicable, maternal, perinatal and nutritional conditions' and 'injuries' are relevant to this study. The report on *Medical Certification of Cause of Death 2017* identifies eight leading cause groups of death. Of these, "Other unspecified effects of external causes and certain early complications of trauma" are relevant to this study. This category includes International Classification of Disease 10 codes T33-T35 and T66-T73. This is relevant to this study because starvation (T73.0) is included in this category.

The *Sample Registration Survey Statistical Report 2018* overviews the percentage of deaths attended by qualified professionals and untrained functionaries. The *Accidental Deaths and Suicides in India 2018* report categorises the fatalities into three categories: accidents, suicides, and injuries. According to this report, 'epidemic' (e.g. COVID-19) falls under the category of 'accidental deaths' caused by the forces of nature. "Other causes" include traffic accidents, which consist of three subcategories: (i) road accidents, (ii) railway crossing accidents, and (iii) railway accidents. These categories and subcategories are relevant to this study. The category of Crimes information such as custodial deaths, complaints against the police injuries and casualties of police and civilians and causes of deaths from "person not on remand" and "person in remand" are relevant to this study because the causes include (i) suicide, and (ii) injuries sustained due to physical assault by police.

The broad categories of causes of death, fatalities, and crimes were used to record the avoidable disaster deaths that were occurring in the first phase of COVID-19 and economic lockdown. These broad categories and sub-categories also act as reference points to interpret the findings of the content review undertaken for this study in 2020, and the data is presented in the subsequent chapters.

References

Director's Desk. (2022). *Director's message*. National Crime Records Bureau. Ministry of Home Affairs, Government of India. Available online https://ncrb.gov.in/en/directors-desk. Accessed on September 3, 2022.

NCRB (National Crime Records Bureau). (2019a). *Accidental deaths and suicides in India 2018*. Ministry of Home Affairs, Government of India. Available at: https://www.ncrb.gov.in/accidental-deaths-suicides-in-india-year-wise.html?year=2018&keyword=. Accessed on June 15, 2020.

NCRB (National Crime Records Bureau). (2019b). *Crime in India—2018*. Ministry of Home Affairs, Government of India. Available at: https://www.ncrb.gov.in/crime-in-india-year-wise.html?year=2018&keyword=. Accessed on June 15 2020.

ORGI (Office of the Registrar General of India). (n.d.). *Presentation on causes of death in India 2010–2013*. Available at: http://censusindia.gov.in/vital_statistics/VA_Report_2010-2013_Release.xlsx. Accessed on June 15, 2020.

ORGI (Office of the Registrar General of India). (2019). *Report on the medical certification of cause of death 2017*. Available at: https://censusindia.gov.in/nada/index.php/catalog/40070. Accessed on October 24, 2024.
ORGI (Office of the Registrar General of India). (2020a). *SRS bulletin*. Available at: https://census india.gov.in/vital_statistics/SRS_Bulletins/SRS%20Bulletin_2018.pdf. Accessed on June 15, 2020.
ORGI (Office of the Registrar General of India). (2020b). *Sample registration system statistical report 2018*. Available at: https://censusindia.gov.in/nada/index.php/catalog/44374. Accessed on October 22, 2024.
ORGI (Office of the Registrar General of India). (2021a). *About the ministry*. Available at: https://censusindia.gov.in/census.website/about/mha. Accessed on September 3, 2022.
ORGI (Office of the Registrar General of India). (2021b) *ORGI functions*. Ministry of Home Affairs, Government of India. Available online https://censusindia.gov.in/census.website/node/297. Accessed on September 3, 2022.
ORGI (Office of the Registrar General of India). (2014a). *Extending the scheme of Medical Certification of Cause of Death (MCCD) to all medical institutions-regarding*. New Delhi. No. 1/1/2014-VS (MCCD)/3143-3210. Available at: https://censusindia.gov.in/2011-Documents/mccd_Report1/Circular2.pdf. Accessed on June 21, 2020.
ORGI (Office of the Registrar General of India). (2014b) *Special registration system: Causes of death statistics 2007–2013*. Available at: SRS_COD-STATISTICS_2007-2013 (1).pdf. Accessed on October 22, 2024.
ORGI (Office of the Registrar General of India). (2014c) *Special registration system: Cause of death statistics 2010–2013*. Available at: file:///C:/Users/nibed/Downloads/SRS_COD-STATISTICS_2010–2013.pdf. Accessed on October 22, 2024.
PARIS21 and OECD/DCD. (2013). *Counting down poverty: The role of statistics in world development*. Analytical report. www.paris21.org. Available at: https://paris21.org/sites/default/files/2532.pdf. Accessed on October 22, 2024.
Whitty, C. (2017). *The eradication of infectious disease*. Gresham College. Available at: https://www.gresham.ac.uk/watch-now/eradication-infectious-diseases. Accessed on March 17, 2023.
WHO (World Health Organization). (2017). *World health statistics 2017*. Available at: https://apps.who.int/iris/bitstream/handle/10665/255336/9789241565486-eng.pdf?sequence=1. Accessed on July 23, 2020.

Chapter 5
Risk Governance for COVID-19 in India

> *With 50% of the knowledge, We have to make 100% of the decisions*
> —Mr. Mark Rutte, Prime Minister of the Netherlands (Quoted in Beaney et al., 2020, p. 330)

Abstract This chapter presents the novel disaster risk governance (DRG) approach adopted by the Government of India during COVID-19 under the remit of the National Disaster Management Act 2005. DRG is one of the components of the avoidable disaster deaths approach. Lockdowns and Unlocks were the critical strategies to contain the spread of COVID-19. The components of DRG, communication (C), coordination (C), collaboration (C) and a combination of all 3Cs are studied in the context of these strategies. Data for these components were gathered from the websites of the Ministry of Health and Family Welfare, The Press Information Bureau, the National Disaster Management Authority and The Ministry of Home Affairs from January to December 2020. Fifteen examples of communication, coordination, collaboration and 3Cs are provided. The limitations of data are also discussed. The findings set the context to understand the circumstances surrounding avoidable disaster deaths in the first phase of COVID-19.

Keywords Avoidable disaster deaths · Disaster risk governance · Coordination · Collaboration · Communication · 3Cs · Lockdown · Unlock · National Disaster Management Act 2005

Introduction

Disaster deaths are conceived as avoidable and unavoidable. The previous chapter presented data on unavoidable deaths. This chapter focuses on avoidable disaster deaths. Avoidable disaster deaths are conceived as preventable, amenable and disaster risk governance-related (discussed in Chap. 3). This chapter focuses on the novel 'disaster risk governance' (DRG) undertaken by the Government of India during the first wave of COVID-19. This chapter aims to contextualise how this governance

N. S. Ray-Bennett, *Avoiding Disaster Deaths*,
https://doi.org/10.1007/978-3-031-81300-9_5

structure may have led to reduced disaster deaths as well as contributed to preventable, amenable, and governance-related avoidable deaths.

The spread of COVID-19, which was declared a pandemic on 11 March 2020, spread at an unprecedented speed and scale in less than a year. This compelled national governments, policy-makers, and healthcare professionals to make crucial decisions based on limited evidence in an uncertain time. This is explained in the fitting words of the then Dutch Prime Minister Mr. Mark Rutte, "with 50% of the knowledge we have to make 100% of the decisions" (Beaney et al., 2020, p. 330).

The World Health Organization (WHO) rolled out several prevention interventions to help curb the spread of the virus, such as hand hygiene, respiratory etiquette, social distancing, crowd control, and lockdown—as mentioned in Chap. 1. National governments, in a very short period 'with 50% of the knowledge' as succinctly put by Mark Rutte, implemented several of these and, most significantly, economic lockdowns. According to Klein et al. (2020) India was predicted to experience the impact of this biological disaster disproportionately affecting around 300–400 million Indians by July 2020, with a likelihood of recurrence in November/December 2020.

To curb the spread of the virus in the most populous country in the world, on 24 March 2020, the National Disaster Management Authority (NDMA) , within the remit of Section 6(2)(i) of the 'Disaster Management Act, 2005' invoked this Act and directed all Ministries/Departments of the Government of India and State Governments to adopt measures to enforce social distancing in India (NDMA, 2020). On the same day, the Ministry of Home Affairs released an order calling for a lockdown across the country for 21 days (Ministry of Home Affairs, 2020a). Subsequently, the government announced another three lockdowns (discussed below). In this context, it is essential to understand this novel disaster risk governance adopted to enforce these lockdowns and how they may have contributed to avoidable disaster deaths.

Effective disaster risk governance is vital to save lives during disasters (Alam & Ray-Bennett, 2021; Mizutori, 2020; Ray-Bennett, 2018). Similarly, a lack of or centralised disaster risk governance can be equally detrimental to saving lives during disasters. Therefore, it is paramount to understand the causes and circumstances of avoidable disaster deaths in India's first phase of COVID-19 by enforcing the governance surrounding lockdowns. Twenty-nine real cases of death are drawn to illustrate this in the next chapter.

First, this chapter introduces some basic concepts of lockdown followed by the different phases that were introduced in India. Second, this chapter presents the disaster risk governance undertaken by the Government of India to enforce the various stages of lockdown along with measures introduced to reduce preventable and amenable avoidable deaths. This is demonstrated through four components of disaster risk governance: communication (C), collaboration (C), coordination (C), and a combination of all 3Cs. This chapter also presents India's everyday governance and administrative structures to contextualise disaster risk governance. Finally, a brief discussion of the limitations of the data is also presented.

Lockdowns: Policy and Practice

A key concept behind the idea of a lockdown is quarantine, defined as a measure that "separates and restricts the movement of people who were exposed to a contagious disease to see if they become sick [...] and helps prevent the spread of disease that can occur before a person knows they are sick or if they are infected with the virus without feeling symptoms" (CDC, 2017). Some of the earliest references to the practice of isolation strategies to prevent the spread of disease are found in the Old Testament (Mackowiak & Sehdev, 2002). Chapter 13 of Leviticus states that anyone suffering from leprosy must live outside the camp, away from others (Mackowiak & Sehdev, 2002). The modern concept of preventative quarantine, strictly related to the plague, is believed to have originated in the fourteenth century as a measure to protect inhabitants of coastal cities from plague epidemics and was first introduced in 1377 in Dubrovnik, Croatia, when the Rector officially issued the *trentina* (an Italian word derived from '*trenta*', the number 30) which was a 30-day isolation period (Gensini et al., 2004). Similarly, in Italy, ships arriving from infected ports in Venice were required to sit at anchor for 40 days or *quaranta giorni*, before they were allowed to land. This is considered the origin of the word quarantine. It is unknown why 40 days were chosen as the duration of isolation. It is suggested that it may have been associated with Hippocrates' theories regarding acute illnesses or that it may be connected to Pythagoras' theory of numbers (Tognotti, 2013).

Since medicines were ineffective against the plague, preventing contact with infected persons and contaminated objects was the only way forward (Tognotti, 2013). Thus, in Italy during the plague epidemic of 1347–1352, some city-states prevented strangers, especially merchants and some minority groups and those suffering from leprosy, from entering their cities (Tognotti, 2013). Armed guards also upheld a sanitary barrier along transit routes as well as at access points to cities. The use of makeshift camps separated healthy and infected persons. "Implementation of these measures required rapid, firm action by authorities, including prompt mobilization of repressive police forces" (Tognotti, 2013, p. 255).

Cholera spread in the nineteenth century, reaching Europe in 1830 and America in 1832 in the context of increasing globalisation (Tognotti, 2013). The first wave of the cholera epidemic saw the adoption of the same public health strategies employed during the plague. New *lazarettos*, or isolation hospitals for those suffering from infectious diseases, were planned at ports; ships arriving from regions affected by cholera were barred entry at European ports; travellers who had contact with infected persons were quarantined and patients of cholera were forced into hospitals and marginalised sections of the population were kept away from cities in 1836. Health officials considered beggars and prostitutes the carriers of the disease and, thus, disallowed their free movement to protect the healthy population in cities (Berridge, 2016; Tognotti, 2013). Such interventions were unknown during peacetime and generated widespread resentment and fear among the people (Tognotti, 2013).

Three waves of an influenza pandemic in the twentieth century were noteworthy. Between 1918 and 1919, public health officials around the globe attempted various disease-containment strategies to stop the spread of the disease, including closing down academic institutions and suspending public gatherings of any kind (Tognotti, 2013). Physicians encouraged respiratory hygiene and social distancing. However, the measures were uncoordinated and implemented late, especially in regions severely affected by the First World War, where the movement of soldiers facilitated the spread of influenza (Tognotti, 2013).

Epidemics in the late 20th and early twenty-first centuries include the Asian Flu Pandemic of 1957, the SARS outbreak of 2003, the influenza A (H1N1 Swine Flu) pandemic, MERS-CoV of 2013, and the Ebola virus outbreak of 2014. The case of Middle East Respiratory Syndrome (MERS) in 2013 in the Republic of Korea is noteworthy. MERS is an illness caused by a virus (more specifically, a coronavirus) called the Middle East Respiratory Syndrome Coronavirus (MERS-CoV) (CDC, 2019). Most MERS patients develop a severe respiratory illness with symptoms of fever, cough and shortness of breath. About 3 or 4 out of every 10 patients reported with MERS have died (CDC, 2019). The disease was first reported in Saudi Arabia in September 2012 and was linked to travel to, or residence in, countries in and near the Arabian Peninsula. The largest known outbreak of MERS outside the Arabian Peninsula occurred in the Republic of Korea in 2015. The outbreak was associated with a Korean traveller returning from the Arabian Peninsula (CDC, 2019). This was one of the most significant outbreaks outside of the Middle East. By the end of the outbreak in July 2015, 186 laboratory-confirmed cases (185 in the Republic of Korea and 1 in China) and 38 deaths were recorded (WHO, 2015).

Intense public health measures, including contact tracing, quarantine and isolation of all contacts and suspected cases, and infection prevention and control, brought MERS-CoV under control in the Republic of Korea (WHO, 2015). Seoul's model of mass testing to slow down the spread and contact tracing of a patient and others who have been exposed through the patient were two novel methods that will become quintessential in not just flattening the curve of COVID-19 but also to 'crunch the curve' without a lockdown in 2020 (Sridhar, 2021). Thanks to the new legislation passed post-MERS in the Republic of Korea to acquire individuals' GPS data from their phones as well as quick access to credit card details and closed-circuit television. In June 2020, the KI-Pass, including a QR code, was introduced to help contact tracing efforts further after outbreaks at several high-risk locations (Sridhar, 2021).

Lockdown in India

India, the second most populous country in the world (whilst writing this book), and home to more than 100 million poor, resorted to a "crude response mechanism to COVID-19—the lockdown" (Sridhar, 2021, p. 221) due to fear that unrestricted movement might result in a torrent of COVID-19 cases, overwhelming the entire health care system (Busch et al., 2020, quoted in Thayer et al., 2021, p. 621). The objective of the lockdown was to contain the virus [not to reduce avoidable disaster deaths], reduce the rate and number of infections and thereby flatten the curve of the COVID-19 pandemic.

From March to June 2020, Indians experienced five Lockdowns (1.0–5.0) and two Unlocks (1.2 and 2.0), the last two of which coincided with the Lockdowns (4.0 and 5.0) (see Table 5.1).

Before the strictest Lockdown 1.0, a Junta Curfew was announced, lasting 14 h. Barely within four hours' notice, India's Prime Minister announced the first Lockdown on 24 March 2020 (Pandey, 2020) as the country reported 598 COVID-19 cases with 12 deaths (Ray et al., 2020; quoted in Thayer et al., 2021). According to BBC News (2021), India had a confirmed caseload of just 519 and nine deaths at the time. Although the number of cases and deaths vary from one source to another during the announcement of Lockdown 1.0, it is useful to note that the numbers were low and that it was announced in haste, and many argue without consultation with the opposition parties (Pandey, 2020; Sen, 2020). The impact of the lockdown on the public is summarised in the next chapter.

Lockdown 1.0 was for 21 days, and the BI India Bureau (2020) defined lockdown as:

> an emergency protocol implemented by the authorities that prevents people from leaving a given area. A full lockdown will mean that the people in the given area must stay where they are and must not exit or enter a building or given area (BI India Bureau, 2020).

Table 5.1 Lockdown and unlock stages and characteristics

Lockdown stages	Lockdown characteristics		
	Dates	No. of hours/ days	Restriction level
Pre lockdown	February 2020		India closes its borders
Junta curfew	22 March 2020	14 h	Voluntary restriction
Lockdown 1.0	25 March–14 April 2020	21 days	Implemented by the Indian Prime Minister Complete restriction
Lockdown 2.0	15 April–3 May 2020	19 days	Complete restriction
Lockdown 3.0	4 May–17 May 2020	15 days	Extension of strict national lockdown—with some relaxations
Lockdown 4.0	18 May–31 May 2020	14 days	Extension of lockdown for most metropolitan cities, red and orange zones with phased reopening plans for green zones Further relaxation
Lockdown 5.0/ Unlock 1.0	1 June–30 June 2020	30 days	Phased reopening plans in certain states. Full relaxation
Unlock 2.0	8 June 2020 onwards		Plans to open schools and colleges. Full relaxation

Sources The author compiled the information from Venkata-Subramani & Roman (2020) and Thayer et al. (2021)

A lockdown can be both a preventive and an emergency strategy for protecting the lives of the vulnerable or persons at risk (Grover et al., 2020). On 24 March 2020, the guidelines released by the Government of India stated that barring specified exceptions (available within the guidelines), all offices, commercial, private and industrial establishments, educational, training and research institutions and places of worship were to be closed for the lockdown period. Hospitality and all transport services were suspended, and all social, political, sports, academic and other institutions were barred. A maximum of 20 persons were permitted to attend a funeral. All persons who arrived in India after 15 February 2020, as well as those directed by healthcare personnel to do so, were required to remain under quarantine, failing which they would be liable for legal action under the Section 188 of the Indian Penal Code permitted (Ministry of Home Affairs, 2020b). However, hospitals, veterinary hospitals, pharmacies, chemists, ambulances, nursing homes and all related medical establishments were to remain functional. Transportation for all medical personnel, para-medical staff and other hospital support services was permitted (Ministry of Home Affairs, 2020b). The Ministry of Home Affairs also invoked provisions of the Essential Commodities Act 1955[1] (currently known as The Essential Commodities (Amendment) Ordinance, 2020) to ensure the availability of essential goods (PRSindia.org, 2020). The Act provides measures such as 'fixing of stock limits, capping prices, enhancing production, an inspection of dealers accounts and other such actions' (PIB, 2020a, p. 1).

On 14 April 2020, the National Disaster Management Authority (NDMA, 2020) "considering the fact that strict social distancing measures need to be implemented for a further period to contain the spread of COVID19" directed the Central and State Governments to continue the same measures until 3 May 2020. Thus, at the end of the initial 21-day lockdown period, Lockdown 1.0 was extended and was called Lockdown 2.0. The Lockdown 2.0 was for 19 days and was strict in implementing the measures. After the second Lockdown, Lockdown 3.0 was announced for 15 days, Lockdown 4.0 was for another 14 days and Lockdown 5.0 was for 30 days. In total, 99 days of lockdown were observed in the first wave of COVID-19. The unlocking began in phases starting on 31 May 2020 along with a rapid scale-up of testing, contact tracing, and isolation, followed by the first stage of relaxing movement restriction starting 1 June 2020 (Thayer et al., 2021).

The relaxing restrictions were called Unlock 1.0, one that eased almost all restrictions other than the continued closure of schools and limiting the number of persons in a social gathering. However, in the six states with the largest number of cases since

[1] The Essential Commodities Act 1955 (p. 2) is "to provide, in the interest of the general public, for the control of the production, supply and distribution of, and trade and commerce, in certain commodities". This Act was amended in 2020 and is currently known as The Essential Commodities (Amendment) Ordinance, 2020 (PRSindia.org, 2020). This Act empowers the Central Government to designate certain commodities (such as food items, fertilisers and petroleum products) as essential commodities. The Ordinance provides that the Central Government may regulate the supply of certain food items including cereals, pulses, potatoes, onions, edible seeds, and oils only under extraordinary circumstances. This includes: (i) war, (ii) famine, (iii) extraordinary price rise; and (iv) natural calamity of grave nature (PRSindia.org, 2020).

May and into the first week of June (Punjab, Madhya Pradesh, Maharashtra, Telangana, Tamil Nadu and West Bengal), the State Governments decided to continue with movement restrictions imposed during Lockdown 4.0. These restrictions were eventually relaxed from 8 June 2020 onwards and were labelled Unlock 2.0 (Thayer et al., 2021).

Although there are a lot of criticisms about the lockdown, a dearth of studies confirm whether the lockdown helped contain the virus, reduce the rate and number of infections, and flatten the curve of the COVID-19 pandemic. Thayer et al.'s. (2021) study in this context is an exception. Using COVID-19 Community Mobility Reports published by Google to assess the effect of the lockdown policy on human mobility and analysed by time series from Lockdown 1.0 and through to 5.0, the authors found a significant reduction in the rate of increase in new COVID-19 cases during Lockdown 1.0 and then Lockdown 4.0, with no significant rebound increase in this rate during the subsequent easing of the lockdown policy (Thayer et al.'s., 2021). Other than Lockdown 1.0, there was no significant increase in the level of COVID-19 incidence. This indicates that the number of reported incident cases did not return to the pre-lockdown trend or the exponential growth pattern associated with the spread of COVID-19 sans prophylactic measures (Thayer et al., 2021). The substantially large decline in the trend of daily incidence that occurred after Lockdown 1.0 may be due to **more vigorous enforcement** [my emphasis] of lockdown measures during this stage than in subsequent stages (Thayer et al., 2021). India's Prime Minister, Mr. Narendra Modi, quickly declared success over the coronavirus, saying, "India was in the endgame of the pandemic" (Sridhar, 2021, p. 231). However, this will turn out to be a premature declaration because the second wave of COVID-19 or the Delta variant, in spring 2021 will take the Indians and the world once again by surprise by a sheer lack of preparedness, the number of deaths and unending pyres of dead bodies burning day and night across Indian villages, towns and cities. The in-depth discussion of the second wave of COVID-19 is beyond the remit of this study.

Disaster Risk Governance for the Lockdown and Unlocks

Enforcing 99 days of five Lockdowns and two Unlockings in one of the most populous countries on this planet is not easy. Critics called the enforcement "one of the strictest lockdowns early in the pandemic due to fear that unrestricted movement might result in a torrent of COVID-19 cases, overwhelming the entire healthcare system" (Busch et al., 2020, quoted in Thayer et al., 2021, p. 621). It is argued that this enforcement was possible because of the administrative structures and complex web of actors needed to implement this, which is explained through the framework of disaster risk governance.

The Government of India invoked the National Disaster Management Act 2005[2] on 24 March 2020—as mentioned earlier—and directed all Ministries/Departments of the Government of India and State Governments to adopt measures to enforce social distancing. On the same day, the Ministry of Health and Family Welfare announced the first lockdown, and the State Governments were directed to invoke the Epidemic Diseases Act 1897[3] and use Section 188 of the Indian Penal Code to implement quarantine. The Ministry of Home Affairs activated the Essential Commodities Act 1955 to ensure the availability of essential goods. The enforcement of these Acts was a seismic decision.

The Disaster Management Act of 2005 empowered the Central government, and the Epidemic Diseases Act of 1897 and Section 188 of the Indian Penal Code empowered the State governments for rapid decision-making. Following these, on 27 March 2020, the Ministry of Health and Family Welfare launched the 'India COVID-19 Emergency Response and Health Systems Preparedness Project P173836'. On 29 March 2020, the Prime Minister of India set up '11 Empowered Groups of Officers' headed by the Chairmanship of the Principal Secretary for planning and ensuring quick implementation of the decision taken to check the spread of COVID-19 (Gupta and Dhingra, 2020; PIB, 2020a). Furthermore, the Prime Minister formed another Committee comprising 15 Union Ministers assigned to 15 districts each and tasked to ensure no disruption in the supply of essential commodities (Gupta and Dhingra, 2020). The enforcement of these legislations by various governmental actors and committees at different levels is discussed further through the four components of disaster risk governance.

---

[2] *The Disaster Management Act 2005* extends to the whole of India. "It can come into force on such date as the Central Government may, by notification in the official Gazette appoint; and different dates may be appointed for different provisions of this Act and for different States, and any reference to commencement in any provision of this Act in relation to any State shall be construed as a reference to the commencement of that provision in that State" (Vishwanathan, 2005).

[3] *The Epidemic Diseases Act 1897* is to "provide for better prevention of the spread of Dangerous Epidemic Disease" (MoHFW, 1957, p. 1). On 22 April 2020, The Ministry of Law and Justice amended this Act to 'The Epidemic Diseases (Amendment) Ordinance, 2020' (Government of India Press, 2020). This decision was taken in the first wave of COVID-19 by the President whilst 'Parliament was not in session' but the circumstances existed to "render it necessary for him to take immediate actions" (Government of India Press, 2020, p. 1). The Amendment introduces (a) "act of violence" committed by any person against a health care service personnel serving during an epidemic shall be punishable, and emphasises the role of the Central Government by introducing the following paragraph: "the Central Government may take such measures, as it deems fit and prescribe regulations for the inspection of any bus or train or goods vehicle or ship or vessel or aircraft leaving or arriving at any land or port or aerodrome, [...] and for such detention thereof, or of any person [...], as may be necessary" (Government of India Press, 2020, p. 3).

Components of Disaster Risk Governance

The four selective components of disaster risk governance for this study are coordination (C), communication (C), and collaboration (C) (Alam & Ray-Bennett, 2021; Ray-Bennett, 2017, 2018). Press releases, memos, advisory guidelines, official letters and reports (documentary evidence) published through the websites of the Ministry of Health and Family Welfare, The Press Information Bureau, the National Disaster Management Authority and The Ministry of Home Affairs were reviewed from January to December 2020 to understand the four components of disaster risk governance. Pre-lockdown months (January and February 2020) were also included to review the disaster risk governance response of the Government of India. Secondary literature was also examined to complement the documentary evidence and understand India's everyday governance and administrative structure. The everyday governance structure and processes have been pivotal for successfully implementing the five lockdowns and two unlockings. The governance and administrative structure of India are discussed briefly below.

Governance Structure of India

The Indian Government consists of 28 States and eight Union Territories. India is neither completely 'federal' nor is it 'unitary' in character. It is thus considered 'quasi-federal'. "It can be better phrased as '*federation sui generis*' or federation of its own kind" (Patna University, n.d, p. 1). There are three tiers of government: (i) the Union/Central Government functioning at the national level; (ii) the State Governments functioning at the state level; and (iii) Panchayats and Municipalities (local government) functioning at the local level (NCERT, 2020). This multilevel governance is illustrated in Fig. 5.1. Each tier has its own jurisdiction, provided in the Indian Constitution, even though they govern the same citizens. The Union/Central Government has three branches (*executive, legislative and judicial*) with different responsibilities but are also interdependent. The *executive branch* comprises the President, Vice President, and a Council of Ministers, led by the Prime Minister. The *legislative branch* consists of two houses of parliament—the lower house, or Lok Sabha (House of the People), and the upper house, or Rajya Sabha (Council of States). The President of India is considered part of the parliament. At the apex of the *judicial branch* is the Supreme Court, whose decisions are binding on the higher and lower courts of the State Governments (Britannica, 2021).

The State Governments function similarly to the way the Union or Central Government does at the national level (Bhan & Rohatgi, 2021). The State Executive of the Government of India consists of the Governor and a Council of Ministers led by the Chief Minister. Most States have unicameral legislatures (with one house of the legislature), while six states have bicameral legislatures (with two houses of the

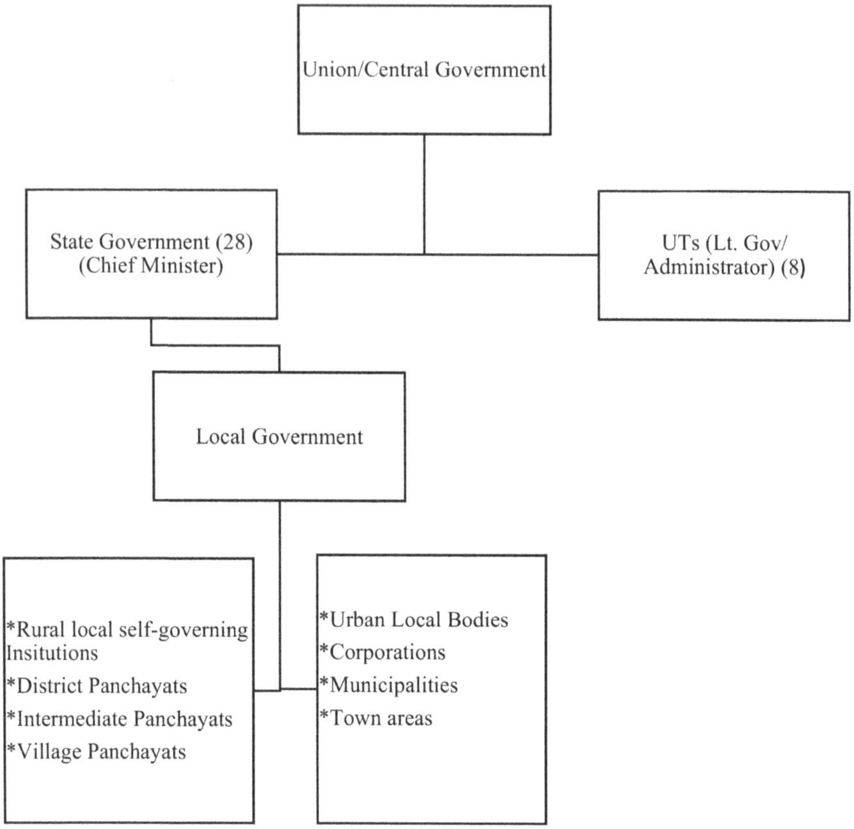

Fig. 5.1 Multilevel Governance in India. *Source* Author

legislature). Some Union Territories, such as Delhi and Puducherry, have legislatures, while others, such as Chandigarh and Daman and Diu, do not. Further, each Indian State has one High Court, although some High Courts have jurisdiction over multiple States and Union Territories (Bhan & Rohatgi, 2021). Indian States are further subdivided into districts, sub-districts, blocks (intermediate) and villages. In urban areas, the subdivision includes urban local bodies, corporations, municipalities, and town areas. Each Indian State has a High Court and each district has a District Court (Bhan & Rohatgi, 2021).

The Governor, as the representative of the President, is the Head of the Executive in the State Governments. The system of government in States closely resembles that of the Union/Central. The President administers Union Territories through an Administrator appointed by him/her (Know India, 2021). Thus, Indian States have their own elected government, where the Governor acts as the representative of the President. Meanwhile, a Union Territory is ruled by the Union/Central Government through a Lieutenant Governor or Administrator appointed by the Union/Central Government.

The three-fold distribution of legislative powers between the Union and State Governments is provided in the Indian Constitution, which contains three lists (NCERT, 2015a): **Union List**: This list includes subjects like defence and foreign affairs and the Union Government alone can make laws pertaining to this list. **State List**: This list includes subjects of State and local importance, such as agriculture, trade and commerce, and police, and the State Government can make laws pertaining to this list. **Concurrent List**: This list includes subjects like forests, trade, and adoption and succession, and both Union and State Governments can make laws pertaining to this list. In the case of disagreements, the decision of the Union Government prevails. Article 370 of the Indian Constitution states that a State's concurrence is required to make any laws in matters mentioned in the Union and Concurrent lists (NCERT, 2015a). However, it was noted that the Central Government passed several orders about the State list during the pandemic.

Administrative Structure

The administrative structure for the Union/Central Government includes the (i) President, (ii) Vice President and (iii) Prime Minister, who leads his (iv) Council of Ministers to aid and advise the President in the exercise of his duties (Embassy of India, 2021). They form the political executive (Embassy of India, 2021). The permanent executive includes civil servants, who are appointed long-term. The political executive is non-permanent as they are subject to elections every four years. The Secretaries, Additional Secretaries and Joint Secretaries are among the top-most members of the permanent executive (Ramabrahmam, nd). The senior civil servant of the Indian Government is the Cabinet Secretary. They are under the direct charge of the Prime Minister and act as the chief coordinator of the Central Government.

The administrative structure of the State Government includes the (i) Governor and (ii) Chief Minister, who leads his/her (iii) Council of Ministers. The President appoints the Governor of a State for a five-year term and holds office (Embassy of India, 2021). India has a prime ministerial form of government. Thus, while the President is the head of the Indian State and is the highest formal authority in the country, the Prime Minister is the head of the Government and exercises all governmental powers (NCERT, 2015b). They make most of the decisions in cabinet meetings (NCERT, 2015b). This was the case during and after the first wave of COVID-19. Also, these existing centralised and decentralised governance and administrative structures became the main conduit for implementing and enforcing lockdowns, unlockings, and advisories generated by the Central Government. This is demonstrated in the next section through the evidence gathered for coordination, communication, collaboration, and combination of 3Cs between the Central and State Governments/Ministries/Departments in implementing the lockdowns, unlocking, enforcing social distancing, reducing the rate and number of infections, and preparing emergency response and health systems for amenable and preventable interventions—among other things.

Table 5.2 Documents reviewed and sample size of each source

	Source reviewed (government portal/website reviewed)	Sample size (numbers of articles/entries entered)
After Accuracy Check (268 entries)	The Ministry of Health and Family Welfare	223
	The Press Information Bureau	43
	National Disaster Management Authority	1
	The Ministry of Home Affairs	1

Findings: Disaster Risk Governance

After reviewing the website and portals of the Ministry of Health and Family Welfare, the Press Information Bureau, the National Disaster Management Authority and the Ministry of Home Affairs for 12 months, 268 pieces of documentary evidence were shortlisted for deeper analysis (see Table 5.2). Of this, the frequency of entries was highest for the website of the Ministry of Health and Family Welfare at 223 entries (83.2%), followed by the Press Information Bureau at 43 (16%) and others at two entries (< 1%).

The amount of documentary evidence posted by these four portals/websites increased sharply first between March and April 2020 (78 and 61, respectively) (before the lockdown was implemented) and slightly again from May to June, after which it saw a gradual fall to five in October and November and increased to 20 in December (see Fig. 5.2). This increase may be explained by the vaccination rollout, a subject on which many of the December 2020 releases are based. This data is not without its limitations. A team of three persons reviewed four Union/ Central Government-run portals/websites manually,[4] every day. We may have overlooked some documents because we did not use any machine-learning tools. Also, we excluded the State Government portals/websites of 35 States and Union Territories, which was beyond the scope of this study. Nonetheless, the purpose of this research is to illustrate the practical application of avoidable disaster deaths, and this was achieved with the successful review of the content review team.

Of the 268 pieces of documentary evidence (see Table 5.3), the number of entries was disaggregated under the following headings of DRG:

(i) 138 were categorised as communication
(ii) 11 were categorised as coordination
(iii) 16 were categorised as collaboration
(iv) 9 were categorised as communication + collaboration
(v) 52 were categorised as communication + coordination
(vi) 31 were categorised as coordination + collaboration
(vii) 11 were categorised as communication + collaboration + collaboration.

[4] We did not use any artificial intelligence tool to mine or track data for the selective portals/websites.

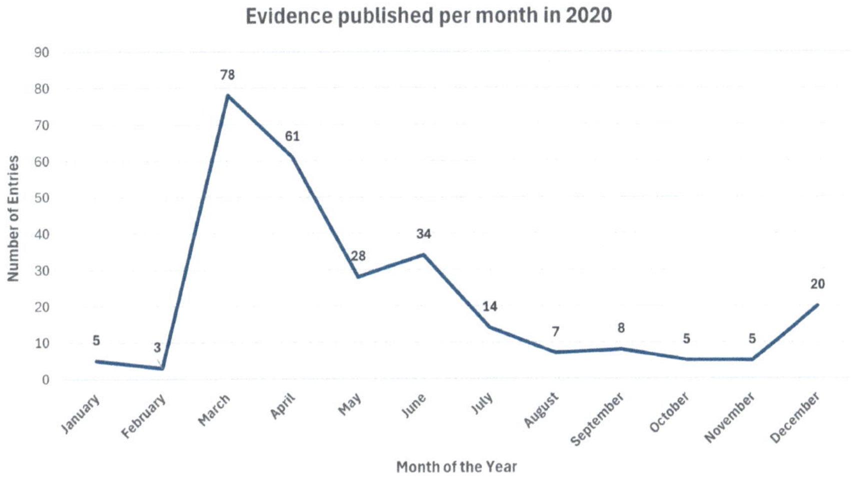

Fig. 5.2 Number of entries

Table 5.3 Number of entries disaggregated by different types of DRG

Total number of accepted data entry January–December 2020	
DRG indicators	Number of accepted entries
Communication	138
Coordination	11
Collaboration	16
Communication + Coordination (combi 1)	52
Coordination + Collaboration (combi 2)	31
Communication + Collaboration (combi 3)	9
Communication + Collaboration + Coordination (combi 4)	11

The highest number of entries is found under the indicator of Communication[5] with 138 entries (51.49%), followed by Communication + Coordination with 52 entries (19.4%) and Coordination + Collaboration with 31 entries (11.56%). The frequency of entries for Communication + Collaboration is the lowest, at nine entries (3.35%). Examples of disaster risk governance are provided below. See Fig. 5.3, which demonstrates different components of disaster risk governance in relation to actors and target audience in the context of COVID-19 in India.

[5] The reason for this high number of entries for communication may be explained by the fact that during a pandemic, information on the spread of the virus, advisories on behavioural health, COVID-19 appropriate behaviours, and other such information is crucial for all citizens to be aware of to prevent and contain the spread. Misinformation and a lack of awareness among citizens, travellers and others can and have led to a rapid spread of the virus in some cases.

Fig. 5.3 Chakra of disaster risk governance. *Source* Author

Communication

About 138 different subjects of communication were found. The communications took place within the government (for example, between the civil servants of the Ministry of Health and Family Welfare and the Ministry of Shipping) and from the Prime Minister to the citizens through press releases, memos, letters and awareness materials. Official websites of The Press Information Bureau, Ministry of Health and Family Welfare, and YouTube were used by the Central Government to communicate with the State Governments, ministries and departments, citizens, travellers, residents, medical professionals, and other groups. *Core information* (discussed in Chap. 1) was communicated through the Ministry of Health and Family Welfare website, YouTube, and official government communications. Here, *core information* includes early warning information, the dissemination of which allows the government to save lives. Some examples of core information include (although not limited to): early travel advisories, warning citizens and travellers against travel, information on protecting oneself during travels, and awareness material on behavioural health. The target audience for the core information included: different actors/responders (such as government officials, airport authorities, employees of quarantine facilities and hotels and medical professionals), citizens, and both actors and citizens. A summary of three examples of communication is provided here.

Example 1*: Novel coronavirus outbreak in China Travel: Advisory to travellers visiting China* (MoHFW, 2020a): this advisory describes the 41 confirmed cases and travellers to China are advised to follow 'simple public health measures at all times.'

Example 2*: The release of a relief package for the poor* (MoHFW, 2020b): This relief package introduced by the Government of India, titled Pradhan Mantri Garib Kalyan Yojana (*aka* the Prime Minister's Scheme for the Welfare of the Poor) was published on 26 March 2020, two days into Lockdown 1.0. The Union Finance and Corporate Affairs Minister announced a relief package Rs 1.70 lakh crore (~USD 234,648) for the poor. The components of the package included: (i) insurance for health workers fighting COVID-19 in government hospitals and health care centres; (ii) Pradhan Mantri Garib Kalyan Anya (food) Yojana; (iii) Benefit to farmers (Rs. 2000 (US$ 24)) to be given in two instalments for 87 million farmers; (iv) Cash transfers under PM Garib Kalyan Yojana; (v). financial support to women's self-help groups; (vi) Other components of Pradhan Mantri Garib Kalyan Package included directions for the State Governments to utilise 'Building and Other Construction Workers Welfare Fund' (MoHFW, 2020b).

Example 3: *Minding your minds during COVID-19* (MoHFW, 2020c): This advisory published on 31 March 2020, focuses on emotional and mental health. Most importantly, it explains why lockdown is essential and what to expect from lockdown and isolation. It offers six ways to handle social isolation, six ways to tackle misinformation, rumours, and theories, five ways to handle emotional problems, and three ways to tackle persons with mental illness—among other things (MoHFW, 2020c).

Coordination

About 11 different subjects of coordination were found. Four can be classified as top-down coordination, and three are horizontal coordination. The top-down coordination was led by the Central Government (led in turn by the Prime Minister), which assigned roles and responsibilities to the various actors/Ministries/States/Departments to carry out and implement interventions to contain the spread of coronavirus, to provide essential services to the public or other crucial tasks. The horizontal coordination was occurring between States and Union Territories and their administrative machinery, various ministries and departments, health care facilities, healthcare professionals and other relevant organisations and institutions. To demonstrate top-down and horizontal coordination, a summary of five examples is provided—three of which are top-down and two are horizontal coordination.

Example 1 Top-Down Coordination:Top-Down Coordination: *Cabinet Secretary letter to all Ministries on COVID-19 management* (MoHFW, 2020d): This is a letter from the Cabinet Secretary addressed to Secretaries of all Ministers/Departments released on 8 March 2020 gives explicit instruction to "step up their efforts and fully mobilize their resources to support the efforts of the Ministry of Health & Family Welfare in preparedness, control and containment measures". The letter requests to 'take expeditious action and mobilize all necessary support in this regard' and identifies specific tasks in an Annexure that need to be undertaken by 16 different Ministries,[6] the Department of Pharmaceuticals, the Directorate General of Foreign Trade, the Department of Science and Technology/Biotechnology and the Indian Council of Medical Research (MoHFW, 2020d).

Example 2 Top-Down Coordination: *PMO reviews efforts of eleven empowered groups towards tackling COVID-19* (PIB, 2020b): This memo published on 10 April 2020 provides some details of one of the periodic reviews that took place at the Prime Minister's Office under the Chairmanship of the Principal Secretary to Prime Minister to monitor the ongoing efforts to counter the impact of the pandemic. The Principal Secretary reviewed the efforts undertaken by the Empowered Groups and expressed satisfaction with the detailed testing protocol and procedure. The memo also mentions that all Chief Secretaries of State were issued instructions to arrange shelter for vulnerable groups like migrants and the homeless. Also, the Centre or Central or Union Government at the national level is in constant touch with States and District levels for monitoring, and rollout of Pradhan Mantri Garib Kalyan Yojana—among other things (PIB, 2020b).

Example 3 Top-Down Coordination: *PM interacts with three teams working on developing and manufacturing COVID-19 vaccine* (PIB, 2020c): this press release published on 30 November 2020 describes the Prime Minister's virtual meetings with Gennova Biopharmaceuticals Ltd Pune, Biological E Ltd Hyderabad and Dr. Reddys Laboratories Ltd Hyderabad—three groups working to develop and manufacture a vaccine for COVID-19. The Prime Minister appreciated the efforts being undertaken by their respective scientists to develop vaccines, asked the companies to contribute suggestions for regulatory processes and logistics, emphasised the importance of informing the general public regarding the vaccine, and advised all the departments to engage with the manufacturers so that the efforts of these manufactures would

[6] The 16 different Ministries are the Ministry of Home Affairs (including National Disaster Management Authority, National Disaster Response Force, Central Armed Police Forces, Bureau of Immigration, Land Port Authority etc.); Ministry of External Affairs; Ministry of Defence; Ministry of Railways; Ministry of Labour; Ministry of Civil Aviation; Ministry of Shipping; Ministry of Consumer Affairs, Food and Public Distribution; Ministry of Information and Broadcasting; Ministry of Panchayat Raj; Ministry of Housing and Urban Development Department; Ministry of Transport; Ministry of Textiles; Ministry of Minority Affairs; and Ministry of Tourism.

have high outreach and impact (PIB, 2020c). This press release is an example of the Prime Minister's effort to coordinate with governmental and non-governmental entities (vaccine developers and manufacturers) to achieve the larger goal of vaccine production for the nation on time.

Example 4 Horizontal Coordination: *Guidelines for social distancing* (MoHFW, 2020e): This letter containing guidelines was released on 13 March 2020 by the Health Secretary to Chief Secretaries of all States/Union Territories and Addl.CS/ Pri. Sec/Secretary Health of all States/Union Territories. It clearly outlines that 'as part of management efforts,' specific social distancing measures, such as quarantine, need to be imposed. The letter also identifies that in the larger community's interest, 'we may have to issue instructions to non-cooperative individuals'. For this, the Epidemic Act of 1897, Disaster Management Act and various other provisions under the Indian Penal Code and Criminal Procedure Act etc. can be used to manage infectious diseases (MoHFW, 2020e). Furthermore, the letter mentions "As states get empowered under the Epidemic Disease Act and other provisions, in our public narrative and press briefings, we have to make it very clear that there is no epidemic of COVID-19 in the country. All such actions are primarily legal empowerment for the preventive action that requires to be taken" (MoHFW, 2020e).

Although the tone of this letter is prescriptive and top-down, it assures the Chief Secretaries of all States and Union Territories that they can use these Acts mentioned above and provisions to eradicate COVID-19 at the State and local levels. The letter also encloses an advisory of 15 interventions. To implement these interventions, it was essential to create horizontal coordination at the state, district, sub-district, and village levels.

Example 5 Horizontal Coordination*: Guidelines for social distancing* (MoHFW, 2020f): The Health Secretary released this letter on 19 March 2020 and emphasised the need for "uniform implementation of guidelines on social distancing". This is because some States/Union Territories were not implementing the guidelines uniformly, so the Health Secretary urged them to do so. The letter outlines 12 social distancing measures for all the States/Union Territories to implement (MoHFW, 2020f). These 12 measures affect coordination amongst all sectors, citizens, governmental, non-governmental and private organisations and state, district, sub-district and village levels.

Collaboration

About 16 different collaboration subjects were identified, of which three examples are provided below. The collaboration was noted between the Government of India's Ministries and Departments for policy implementation (such as the implementation of lockdown) as well as with the private sector (such as the development of vaccines). The Central Government primarily led collaboration to promote horizontal collaboration and cooperation with different departments, ministries, State Governments and Union Territories administrations to contain the spread of COVID-19, mitigate the crises resulting from the virus and lockdown, and ensure the implementation of policies. A summary of three examples is provided.

***Example 1**: High-level group of ministers reviews current status, and actions for prevention management of COVID-19* (PIB, 2020d): This travel advisory published on 11 March 2020, explains the outcome of a high-level Group of Ministers meeting that took place at Nirman Bhawan under the chairpersonship of the Union Minister of Health and Family Welfare. The meeting was attended by eight ministers including the Minister of Civil Aviation, Minister of External Affairs, Minister of State for Home, Minister of State (I/C), Minister of Shipping, Chemicals and Fertilizers and Minister of State, Health and Family Welfare. The Group of Ministers took eight decisions to promote preventive measures and preparedness for COVID-19 that demand collaboration (PIB, 2020d).

***Example 2**: Environment and Social Commitment Plan* (MoHFW, 2020g): The Environment and Social Commitment Plan for 'India COVID-19 Emergency Response and Health Systems Preparedness Project P173836' funded by the International Bank for Construction and Development was published on 27 March 2020. This project was implemented by the Ministry of Health and Family Welfare, the Indian Council of Medical Research and the National Center for Disease Control (MoHFW, 2020g). This project is an excellent example of horizontal collaboration across different health agencies and units nationwide.

***Example 3**: Standard operating procedure for passenger movement post disembarkation* (MoHFW, 2020h): This standard operating procedure published by the Ministry of Civil Aviation for Airport Authorities outlines 15 procedures for 'passenger movement post disembarkation'. The procedure outlines the roles of different actors in collaboration to implement these procedures, including airline staff; health professionals manning the Health Counters for thermal screening; Immigration Officers and staff; Central Industrial Security Force personnel; Delhi International Airport staff; and Delhi Police personnel (MoHFW, 2020h). They are required to collaborate to ensure that passengers are correctly screened for COVID-19.

Combi or Combination of 3Cs

Sometimes, a particular piece of evidence did not fit neatly into the communication, collaboration, or coordination component. Since these indicators are not mutually exclusive, a combination of these components was developed to suit specific entries better. About 103 different subjects of a combination of 3Cs were identified, of which 52 are classified as a combination of communication and coordination, 31 as a combination of coordination and collaboration, nine as a combination of communication and cooperation, and 11 as a combination of communication, collaboration, and coordination. A summary of four examples is provided.

Example 1 Combination of Communication and Coordination: *Instructions to all major and minor ports for dealing with novel coronavirus* (MoHFW, 2020i): This letter released by the Directorate General of Shipping, Mumbai, on 20 March 2023 provides 26 instructions to all major and minor ports to deal with the COVID-19 pandemic. This letter is evidence of communication to "all stakeholders" (MoHFW, 2020i, p. 1) that are required to coordinate to implement the instructions. Further, it is also evidence of 'coordination' as the document lists the roles and responsibilities of the "port stakeholders" (MoHFW, 2020j, p. 23), thus acting as the coordinator for the implementation of the instructions. This responsibility for coordination is also assigned to "coordination authorities" and "coordinating officers" in the document (MoHFW, 2020i, p. 23). The stakeholders include the Port Health Officer, Chief Medical Officer, Traffic Manager I, Deputy Conservator/Harbour Master, Ship Agents, Ship Master, Customs Officials, Immigration Officials, Security Agencies, Terminal Operators and Ship Suppliers/Handlers (MoHFW, 2020j).

Example 2 Combination of Coordination + Collaboration: *Visa restrictions related to COVID-19* (MoHFW, 2020j). This letter, released by the Deputy Director of the Bureau of Immigration (Ministry of Home Affairs) on 11 March 2020, provides fresh advice on visa restrictions. The letter was sent to the Additional Secretary of the Ministry of Home Affairs, Joint Secretary of the Ministry of Health and Family Welfare, Joint Secretary of the Cabinet Secretariat, Joint Secretary of the Ministry of External Affairs, Secretary of the Ministry of Tourism, and all Chief Secretaries of State and all Health Secretaries of States. The letter provides the visa restrictions issued by the Ministry of Health and Family Welfare, the implementation of which requires the collaboration of these Ministries and States (MoHFW, 2020j). The letter acts as a coordinating document, as implementing the various restrictions listed requires the interventions of the above Ministries and States.

Example 3 Combination of Communication and Collaboration: *COVID-19 Vaccine rollout* (PIB, 2020d). This press release published on 10 January 2021 by the Press Information Bureau is evidence of communication. At the same time, it is also evidence of the collaboration of the Central Government with the States and Union Territories of India and "all stakeholders" responsible for the COVID-19 vaccine rollout (PIB, 2020e). These stakeholders include health professionals, vaccine developers, the Ministry of Health and Family Welfare and senior civil servants.

Example 4 Combination of communication + coordination + collaboration: *PM interacts with CMs of seven states to review the status and response to COVID-19* (PIB, 2020f). This press release published on 23 September 2020 by the Press Information Bureau updates the general public on the meeting between the Prime Minister and seven Chief Ministers via video conference. As it is a press release, it is considered evidence for communication for the general public. Further, the press release alludes to the Chief Ministers' acknowledgement and appreciation of the leadership of the Prime Minister, thus signalling the role of the Central Government in coordinating COVID-19 prevention and containment measures. Lastly, according to the release, the Prime Minister "stressed the need to increase focus on effective testing, tracing, treatment, surveillance and clear messaging" and also "the importance of ensuring smooth movement of goods and services between States" (PIB, 2020e, p. 3). Through this, the Prime Minister emphasised the need for collaboration among the stakeholders involved in testing, tracing, treatment, surveillance, and messaging, as well as among the Indian States. This collaboration already existed in some cases, and the press release points out the example of the State of Delhi, where "with the combined effort of the people, the State government and the Central Government, the situation was brought under control" (PIB, 2020e, p. 2).

Conclusion

In Chap. 3, it was argued that 'governance' implies a reduction in the role of the state or the government while concurrently recognising the importance of non-state actors in the governance process. In the context of the first wave of COVID-19 under the remit of the National Disaster Management Act 2005, a novel disaster risk governance was adopted—unprecedented in the twenty-first century. This novel disaster risk governance focused on integrating inter-sector and inter-governmental agencies that discharged services to the citizens rather than the involvement of non-state actors (except for the vaccine manufacturers). The evidence provided above, however, is not without its limitations.

The evidence gathered is from the websites and portals of the Central Government, and they over-emphasise the coordination, collaboration, and communication between governmental agencies in a top-down manner. Taking one Indian State as a case study site to observe the implementation of the disaster risk governance components would have provided a deeper insight into understanding the involvement of non-state actors as envisaged by the Sendai Framework. However, due to time and resource constraints, as well as lockdown in the first wave of COVID-19, this was beyond the scope of this study.

Nonetheless, the reviewed data suggest that the State Governments and Departments were implementing the advisories and measures related to the lockdown and unlocking, as well as preventable and amenable measures cascading down from the Central Government. Consequently, the disaster risk governance structure became increasingly centralised. Within that centralised disaster risk governance, the documentary evidence gathered largely from the Ministry of Health and Family Welfare and Press Information Bureau indicates the plethora of actors and organisations involved in the decision-making and implementation.

The imposition of the lockdown meant shutting down nearly all economic activity, which caused a dramatic reduction in the State Governments' revenues. According to Burman (2020) this also increased the State Governments' financial dependence on the Central Government. "Even prior to the lockdown, many States in India had already breached—or came close to breaching—their mandated fiscal deficit limits" (Burman, 2020). This meant further centralisation of some States' power and the erosion of political and fiscal federalism as an unforeseen consequence of the pandemic (Agrawal, 2020; Kashinath, 2020). Furthermore, the National Disaster Management Act 2005 gave the Central Government the power to make policy decisions rapidly and impose restrictions on people to manage this biological disaster (Chauhan, 2020). The Act authorises the National Disaster Management Authority Chairperson, which in turn the Indian Prime Minister, to make decisions to mitigate the disaster impact on matters such as relief for victims and measures to protect the vulnerable sections of society. It also allows the Central Government to "control and restrict the entry of any person into, his movement within and departure from, a vulnerable or affected area" (GoI, 2005).

Thus, the orders and directives—as observed through the four components of disaster risk governance—came in a top-down manner in the form of communications, where decisions had already been made, and ministries, departments, civil servants, and others were expected to assist in implementation, coordination, and collaboration. Having set the context of disaster risk governance, the next chapter brings the circumstances that led to the avoidable disaster deaths in the first wave of COVID-19 to the forefront.

References

Agrawal, P. (2020). COVID-19 and dwindling Indian federalism. *Economic and Political Weekly*, 27 June. Available at: https://www.epw.in/journal/2020/26-27/commentary/covid-19-and-dwindling-indian-federalism.html. Accessed on March 18, 2021.

Alam, E., & Ray-Bennett, N. S. (2021). Disaster risk governance of local level administration in Rangamati, Bangladesh. *International Journal for Disaster Risk Reduction, 59*(2021), 1–13. https://doi.org/10.1016/j.ijdrr.2021.102220

BBC. (2021). India Covid-19: PM Modi 'did not consult' before lockdown. *BBC News*, 29 March. Available online: https://www.bbc.co.uk/news/world-asia-india-56561095. Accessed on March 30, 2021.

Beaney, T., Clarke, J. M., Jain, V., Golestaneh, A. K., Lyons, G., Salman, D., & Majeed, A. (2020). Excess mortality: The gold standard in measuring the impact of COVID-19 worldwide? *Journal of the Royal Society of Medicine, 113*(9), 329–334. Available at: https://www.ncbi.nlm.nih.gov/pmc/articles/PMC7488823/

Berridge, V. (2016). *Public health: A very short description*. Oxford University Press.

Bhan, A., & Rohatgi, M. (2021). *Court structure and hierarchy*. Available at: https://uk.practicallaw.thomsonreuters.com/w-017-5278?transitionType=Default&contextData=(sc.Default)&firstPage=true#co_anchor_a813964. Accessed on March 18, 2021.

BI (Business Insider) India Bureau. (2020). What is lockdown meaning and what to expect from it. *Business Insider India*, 26 March. Available at: https://www.businessinsider.in/india/news/what-is-lockdown-and-what-to-expect-from-the-lockdown/articleshow/74759615.cms. Accessed on September 5, 2020.

Britannica. (2021). *Constitutional structure*. Available at: https://www.britannica.com/place/India/Constitutional-structure#ref46431. Accessed on March 18, 2021.

Burman, A. (2020). *How COVID-19 is changing Indian federalism*. Available at: https://carnegieindia.org/2020/07/28/how-covid-19-is-changing-indian-federalism-pub-82382. Accessed on March 18, 2020.

CDC (Centers for Disease Control and Prevention). (2017). Quarantine and isolation. Available at: https://www.cdc.gov/quarantine/. Accessed on September 5, 2020.

CDC (Centers for Disease Control and Prevention). (2019). About MERS. Available at: https://www.cdc.gov/coronavirus/mers/about/index.html. Accessed on September 18, 2022.

Chauhan, C. (2020). India under Covid-19 lockdown: All about the disaster management law. *Hindustan Times*, 26 March. Available at: https://www.hindustantimes.com/india-news/india-under-covid-19-lockdown-all-about-the-disaster-management-law/story-i7cjfZrUZcbOxamlEOAoPO.html. Accessed on March 10, 2021.

Embassy of India. (2021). *The state structure*. Available at: https://eoi.gov.in/ashgabat/?0774?000#:~:text=State%20executive%20consists%20of%20Governor,holds%20office%20during%20his%20pleasure. Accessed on March 17, 2021.

Gensini, G. F., Yacoub, M. H., & Conti, A. A. (2004). The concept of quarantine in history: From plague to SARS. *Journal of Infection, 49*(4), 257–261. Available at: https://www.journalofinfection.com/action/showPdf?pii=S0163-4453%2804%2900054-4. Accessed on September 5, 2020.

GoI (Government of India). (2005). *The disaster management act, 2005*. Available at: https://www.ndmindia.nic.in/images/The%20Disaster%20Management%20Act,%202005.pdf. Accessed on March 10, 2021.

Government of India Press. (2020). *The Gazette of India*. Available online: https://egazette.nic.in/WriteReadData/2020/219108.pdf. Accessed on April 28, 2023.

Grover, S., Sahoo, S., Mehra, A., Avasthi, A., Tripathi, A., Subramanyan, A., Pattojoshi, A., Rao, G. P., Saha, G., Mishra, K. K., & Chakraborty, K. (2020). Psychological impact of COVID-19 lockdown: An online survey from India. *Indian Journal of Psychiatry, 62*(4), 354. Available at: http://www.indianjpsychiatry.org/article.asp?issn=0019-5545;year=2020;volume=62;issue=4;spage=354;epage=362;aulast=Grover. Accessed on September 11, 2020.

Gupta, M. D., & Dhingra, S. (2020). PM Modi sets up 11 empowered groups as quick response teams to tackle coronavirus outbreak. *The Print*, 29 March. Available at: https://theprint.in/india/governance/pm-modi-sets-up-10-empowered-groups-as-quick-response-teams-to-tackle-coronavirus-outbreak/390849/. Accessed on March 5, 2021.

Kashinath, P. (2020). Centre's adoption of coercive federalism over co-operation with states has hampered response to COVID-19, migrant crisis. *FirstPost*, 17 May. Available at: https://www.firstpost.com/india/centres-adoption-of-coercive-federalism-over-co-operation-with-states-has-hampered-response-to-covid-19-migrant-crisis-8376701.html. Accessed on March 18, 2021.

Klein, E., Lin, G., Tseng, K., Schueller, E., Kapoor, G., & Laxminarayan, R. (2020). *COVID-19 for India updates*. The Center for Disease Dynamics, Economics and Policy. Available at: https://onehealthtrust.org/wp-content/uploads/2020/03/covid19.indiasim.March23-2-eK.pdf. Accessed on June 15, 2020.

Know India. (2021). *States and union territories*. Available at: https://knowindia.gov.in/states-uts/. Accessed on March 17, 2021.

Mackowiak, P. A., & Sehdev, P. S. (2002). The origin of quarantine. *Clinical Infectious Diseases, 35*(9), 1071–1072. Available at: https://academic.oup.com/cid/article-pdf/35/9/1071/943199/35-9-1071.pdf. Accessed on September 5, 2020.

Ministry of Home Affairs. (2020a). *Order (No. 40-3/2020-DM-I(A))*. Available at: https://www.mha.gov.in/sites/default/files/MHAorder%20copy_0.pdf. Accessed on September 5, 2020.

Ministry of Home Affairs. (2020b). *Annexure to Ministry of Home Affairs order No. 40-3/2020-D dated 24.03.2020*. Available at: https://www.mha.gov.in/sites/default/files/MHAorder%20copy_0.pdf. Accessed on September 5, 2020.

Mizutori, M. (2020). Reflections on the Sendai framework for disaster risk reduction: Five years since its adoption. *International Journal of Disaster Risk Science, 11*, 147–151.

MoHFW (Ministry of Health and Family Welfare). (1957). *The epidemic diseases act 1897*. Available at: https://main.mohfw.gov.in/sites/default/files/1625462448.pdf. Accessed on April 28, 2023.

MoHFW (Ministry of Health and Family Welfare). (2020a). *Travel advisory to travellers visiting China.* Available at: https://www.mohfw.gov.in/pdf/TraveladvisorytotravelersvisitingChina17012020.pdf. Accessed on March 12, 2021.

MoHFW (Ministry of Health and Family Welfare). (2020b). *Finance Minister announces Rs 1.70 Lakh Crore relief package under Pradhan Mantri Garib Kalyan Yojana for the poor to help them fight the battle against Corona Virus.* Available at: https://www.mohfw.gov.in/pdf/MoFPMGaribKalyanYojanaPackage.pdf. Accessed on March 17, 2021.

MoHFW (Ministry of Health and Family Welfare). (2020c). *Minding our minds during the COVID-19.* Available at: https://www.mohfw.gov.in/pdf/MindingourmindsduringCoronaeditedat.pdf. Accessed on March 17, 2021.

MoHFW (Ministry of Health and Family Welfare). (2020d). *Cabinet secretary DO letter dated 8th March, 2020 to all Government of India Ministries on COVID-19 management.* Available at: https://www.mohfw.gov.in/pdf/COVIDCabSecLetterGuidelines.pdf. Accessed on March 18, 2021.

MoHFW (Ministry of Health and Family Welfare). (2020e). *DO letter from secretary, DoHFW to state chief secretaries regarding social distancing measures.* Available at: https://www.mohfw.gov.in/pdf/Letterdated2032020toCSs.pdf. Accessed on March 18, 2021.

MoHFW (Ministry of Health and Family Welfare). (2020f). *Advisory on social distancing measures in view of spread of COVID-19 disease.* Letter from Ms Preeti Sudan, Secretary. Letter vide D.O.No. Z-21020/14/2020-PH 16-3-2020

MoHFW (Ministry of Health and Family Welfare). (2020g). *Environmental and Social Commitment Plan (ESCP).* Available at: https://www.mohfw.gov.in/pdf/2EnvironmentalandSocialCommitmentPlanESCP.pdf. Accessed on March 18, 2021.

MoHFW (Ministry of Health and Family Welfare). (2020h). *Standard operating procedure for passenger movement post Disembarkation).* Available at: https://www.mohfw.gov.in/pdf/SOPQuarantineofPassengers.pdf. Accessed on March 18, 2021.

MoHFW (Ministry of Health and Family Welfare). (2020i). *Instructions to all major and minor ports for dealing with novel coronavirus (COVID-19) pandemic.* Available at: https://www.mohfw.gov.in/pdf/DGSOrder04of2020.pdf. Accessed on March 19, 2021.

MoHFW (Ministry of Health and Family Welfare). (2020j). *Visa restrictions related to COVID-19.* Available at: https://www.mohfw.gov.in/pdf/VisarestrictionsrelatedtoCOVID19Ministries.pdf. Accessed on March 19, 2021.

NCERT (National Council of Educational Research and Training). (2015a) *Federalism.* Available at: https://ncert.nic.in/ncerts/l/keps207.pdf. Accessed on March 17, 2021.

NCERT (National Council of Educational Research and Training). (2015b.) *Working of institutions.* Available at: https://ncert.nic.in/ncerts/l/iess405.pdf. Accessed on March 17, 2021.

NCERT (National Council of Educational Research and Training). (2020). *Federalism.* Available at: https://ncert.nic.in/textbook/pdf/jess402.pdf. Accessed on March 17, 2021.

NDMA (National Disaster Management Authority). (2020). *Order (No. 1-29/2020-PP (Pt. II)).* Available at: https://ndma.gov.in/images/covid/ndmaorder240320.pdf. Accessed on September 5, 2020.

Pandey, V. (2020). Coronavirus lockdown: The Indian migrants dying to get home. *BBC News,* 20 May. Available online: https://www.bbc.co.uk/news/world-asia-india-52672764. Accessed on September 17, 2022.

Patna University. (n.d.). *Indian federalism.* Available at: https://www.patnauniversity.ac.in/e-content/social_sciences/geography/Indian%20Federalism.pdf. Accessed on March 17, 2021.

PIB (Press Information Bureau). (2020a). *MHA writes to states to ensure availability of essential goods, by invoking provisions of the Essential Commodities (EC) act 1955, under lockdown to fight COVID-19.* Available at: https://pib.gov.in/PressReleseDetail.aspx?PRID=1612154. Accessed on March 25, 2021.

PIB (Press Information Bureau). (2020b). *PMO reviews efforts of eleven empowered groups towards tackling COVID-19.* Available at: https://www.pib.gov.in/PressReleseDetail.aspx?PRID=1612890. Accessed on March 12, 2021.

PIB (Press Information Bureau). (2020c). *PM interacts with three teams working on developing & manufacturing COVID-19 vaccine*. Available at: https://www.pib.gov.in/PressReleseDetail.aspx?PRID=1677120. Accessed on March 12, 2021.

PIB (Press Information Bureau). (2020d). *High level Group of Ministers reviews current status, and actions for prevention and management of COVID-19*. Available at: https://www.mohfw.gov.in/pdf/HFWnCoVGoM11March2020II.pdf. Accessed on March 12, 2021.

PIB (Press Information Bureau). (2020e). *COVID-19 vaccine rollout: As India gears up for COVID19 vaccine roll out, Centre guides States/UTs on Co-WIN management*. Available at: https://pib.gov.in/PressReleasePage.aspx?PRID=1687421. Accessed on March 12, 2021.

PIB (Press Information Bureau). (2020f). *PM interacts with CMs of seven states to review status and response to COVID-19*. Available at: https://www.pib.gov.in/PressReleseDetail.aspx?PRID=1658407. Accessed on March 19, 2021.

PRSindia.org. (2020). *The essential commodities (Amendment) ordinance, 2020: Ministry consumer affairs and food distribution*. Available online https://prsindia.org/billtrack/the-essential-commodities-amendment-ordinance-2020. Accessed on April 28, 2023.

Ramabrahmam, I (n.d.). *Organizational structure of the Government at the Centre and State*. Department of Political Science, University of Hyderabad. Available at: http://www.mcrhrdi.gov.in/drugs/week1/Oraganizational%20Structure%20of%20the%20Government%20at%20the%20Centre.pdf. Accessed on March 17, 2021.

Ray-Bennett, N. S. (2017). Disasters, deaths and the Sendai goal one: Lessons from Odisha, India. *World Development, 103*(2018), 27–39. https://doi.org/10.1016/j.worlddev.2017.10.003

Ray-Bennett, N. S. (2018). *Avoidable deaths: A systems failure approach to disaster risk management*. Springer Nature.

Sen, A. (2020). Listening as governance. *Culture of Peace News Network*, 21 June. Available at: https://cpnn-world.org/new/?p=20650. Accessed on June 21, 2020.

Sridhar, D. (2021). *Preventable: How a pandemic changed the world and how to stop the next one*. Penguin Random House.

Thayer, M., Hasan, Md. Z., Sankhla, P., & Gupta, S. (2021). An interrupted time series analysis of the lockdown policies in India: A national-level analysis of COVID-19 incidence. *Health Policy and Planning, 36*(2021), 620–629.https://doi.org/10.1093/heapol/czab027

Tognotti, E. (2013). Lessons from the history of quarantine, from plague to influenza A. *Emerging Infectious Diseases, 19*(2), 254. Available at: https://www.ncbi.nlm.nih.gov/pmc/articles/PMC3559034/pdf/12-0312.pdf. Accessed on September 5, 2020.

Venkata-Subramani, M., & Roman, J. (2020). The Coronavirus Response in India–World's Largest Lockdown. *The American Journal of the Medical Sciences*. Available at: https://www.amjmedsci.com/article/S0002-9629(20)30351-7/fulltext. Accessed on September 7, 2020.

Viswanathan, T. K. (2005). The disaster management act 2005. Ministry of Law and Justice. In *The Gazette of India*, No. 64. pp. 1–29. Available online: https://www.ndma.gov.in/sites/default/files/PDF/DM_act2005.pdf

WHO (World Health Organization). (2015). *MERS outbreak in the Republic of Korea*, 2015. Available at: https://www.who.int/westernpacific/emergencies/2015-mers-outbreak. Accessed on April 10,s 2022.

Chapter 6
Avoidable Disaster Deaths in India

> *[...] a calamity would be a case of injustice only if it could have been prevented, and particularly if those who could have undertaken preventive action had failed to try.*
> —Sen (2009)

Abstract This chapter revisits the central questions of this book: How many avoidable disaster deaths occurred from the impact of COVID-19 and the lockdown, and what are the circumstances and causes that led to these deaths? The content of 11 reputable English e-newspapers, three digital news services, and one e-magazine were reviewed from March to December 2020 to capture the number of avoidable disaster deaths. Dynamic strategies were adopted to ascertain the causes and circumstances of death. 520 deaths and 32 different causes were found. Twenty-nine cases of preventable, amenable, and risk-governance-related avoidable disaster deaths are also presented. The economic, humanitarian and psycho-social impact of the lockdown is also discussed. The number of avoidable disaster deaths is not representative but rather illustrative. This demonstrates that the avoidable disaster deaths approach can provide much-needed guidance in data collection and be a practical tool to make sense of disaster deaths to improve disaster management practices and save lives.

Keywords Avoidable disaster deaths · Amenable deaths · COVID-19 · Economic impact · Humanitarian impact · Psycho-social impact · Preventable deaths · Risk governance-related deaths

Introduction

"In the early hours of 12 May—just 200 km from their village—a truck rammed into the tuk-tuk from behind. Sanju and Nandini Yadav[1] died on the spot. Rajan and Nitin Yadavs escaped with minor injuries" (Pandey, 2020). According to the

[1] Sanju Yadav, the wife of Rajan Yadav, died on 12 May. Also died their young daughter, Nandini Yadav (Pandey, 2020). Road accidents caused these deaths.

N. S. Ray-Bennett, *Avoiding Disaster Deaths*,
https://doi.org/10.1007/978-3-031-81300-9_6

deceased husband, "It all ended so quickly [...]. I wish I had gotten the tickets. I wish I had never started the journey [...] I wish I was not poor" (Pandey, 2020).

The circumstances that led to this heartbreaking, avoidable disaster death in the first phase of COVID-19 are telling. This story and others presented in the latter part of this chapter are human tragedies—a tragedy of governance failure, a tragedy of the twenty-first century, a tragedy of our lifetime. The burden of this tragedy was disproportionately carried by the poorest and marginalised sections of society, i.e., the vulnerable groups. This human tragedy could have been mitigated through better planning and preparedness and by centring on the needs of high-risk/vulnerable groups. Albeit, this is in hindsight.

Two central questions were introduced in Chap. 1:

- How many avoidable disaster deaths took place from the impact of COVID-19 and the economic lockdown in India and how do we capture them?
- And what are the circumstances and causes that led to these deaths?

This penultimate chapter answers these questions through the three components of the avoidable disaster deaths approach (amenable, preventable and risk governance-related). It also explains the data collection procedures. By doing so, it brings 29 cases of avoidable disaster deaths to the forefront to explain the causes and circumstances surrounding them. The chapter then concludes by discussing the economic, humanitarian, and psycho-social impact of the lockdown in relation to the number of avoidable disaster deaths.

Data Collection Procedures

The project coordinator and I led a team of 12 content reviewers who reviewed the content of 11 reputable English e-newspapers, three digital news services, and one e-magazine every day for nine months (11 March–31 December 2020) (see Box 6.1). The only exceptions were The Sentinel and The Tribunal, whose review began in July 2020 to represent the northeastern states of India.

The content of these sources was reviewed manually under 12 variables:

Box 6.1 Sources for data

E-newspapers	Digital news services	E-magazine	Others
The Times of India; The Guardian; The New York Times; The New Indian Express; The New Indian Express Odisha; The Hindu; The Tribune; The Sentinel; and Sambad	Bloomberg; BBC News; The Wire; and Scroll	India Today	NDTV, The Print, Hindustan Times, Kalinga TV, The News Minute, The Telegraph, Business Insider, India Blooms, Mumbai Mirror, News Room, Odisha TV, Rural India Online, CBC News

(i) Source
(ii) Day, month and year of release of the article
(iii) URL of the article
(iv) Title of the article
(v) Death-related part of the article
(vi) Age and gender of the deceased
(vii) Type of death (direct/indirect). If direct—reject. If indirect—select. Under indirect deaths use avoidable death indicators (see viii)
(viii) Indicator (amenable/preventable/DRG)
(ix) Cause of death
(x) Cause of death confirmed by (journalist only, journalist using statement of medical officer, journalist using statement of police, content analyst)
(xi) Name of content analyst inputting data
(xii) Day, month and year of input.

Content analysts also reviewed other sources as and when they found deaths-related information and were labelled as 'Others'. In total, 419 articles were reviewed, of which 252 sources were selected. Of the 252 selected articles, 113 (45%) were from The New Indian Express, followed by The Times of India at 28 (11%) and The New Indian Express Odisha and Sambad both at 18 (7%). Therefore, most of the articles were from The New Indian Express. Table 6.1 and Fig. 6.1 show the distribution of articles entered over ten months. The number of entries rose sharply from March to April 2020 and again from April to May (123 highest). After May, the number of relevant articles published per month fell until it reached zero articles in December 2020.

Table 6.1 Distribution of articles entered by month in 2020

The month of publication of an article in the E-newspaper, E-magazine or digital news service	Number of entries
1. March	3
2. April	22
3. May	123
4. June	110
5. July	62
6. August	38
7. September	34
8. October	15
9. November	12
10. December	0

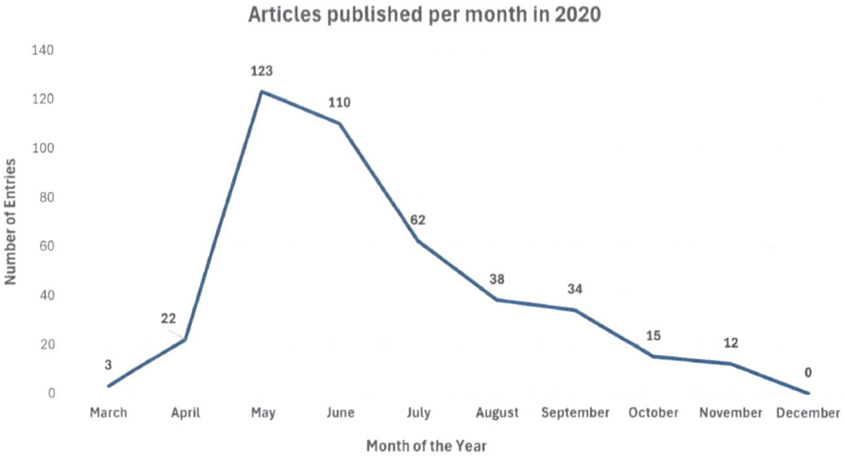

Fig. 6.1 Distribution of articles entered by month in 2020

Findings: Existing and Emergent Causes of Avoidable Disaster Deaths

Three broad categories and indicators of avoidable disaster deaths were introduced in Chap. 3: (i) amenable death, (ii) preventable death, and (iii) disaster risk governance (DRG)-related death. Using these categories and indicators, 56 articles qualified as amenable deaths, 85 as preventable deaths and 111 as DRG-related deaths. Thus, articles on DRG-related deaths were the highest.

Thirty-one broad categories of causes of death were found (see Table 6.2). Although these categories were disaggregated under the three indicators of avoidable disaster deaths, it was noted that some causes and circumstances of deaths do not fit neatly under each type. Therefore, two new categories of avoidable disaster deaths were introduced: "unknown deaths" and "COVID-19 deaths".

"Unknown deaths" as an emergent indicator was found to cross-cut any of the three avoidable death categories (amenable, preventable, disaster risk governance). For example, some deaths are caused by the lack of timely and efficient healthcare in the case of denial of medical care, but the exact cause of death is unknown as no tests were run when the case was reported by the author/journalist. Similarly, migrants walking hundreds of miles collapsed on the road to home, and their exact medical cause of death is unknown. However, it is considered a disaster risk governance-related death as migrants were left with no option but to commence a long journey on foot due to the economic Lockdown 1.0.

Likewise, "COVID-19 deaths" can be in any of the three indicators of avoidable disaster deaths. These are not counted as direct deaths because the patient in question either (i) did not have access to timely and effective health care, thus making it an amenable death; (ii) died because of a failure/lack of public health interventions,

Table 6.2 Cause of avoidable disaster death

Indicators for avoidable disaster deaths	Causes of deaths
Disaster Risk Governance-related deaths	1. Accidental electrocution
	2. Exhaustion
	3. Low hospital bed availability
	4. Low hospital oxygen availability
	5. Low hospital bed availability and medical negligence
	6. Medical negligence
	7. Employer negligence
	8. Medical and employer negligence (combined)
	9. Road or motor vehicle accident
	10. Rail accident
	11. Police brutality
	12. Starvation
	13. Exhaustion and starvation
	14. Prison overcrowding
	15. Forced migration
	16. Poor health and safety
	17. Stigma
	18. Suffocation
Preventable deaths	19. Suicide
	20. Murder
	21. Poisoning
	22. Assault
Amenable deaths	23. Delayed ambulance and low hospital bed availability
	24. Low hospital bed availability and denial of medical care
	25. Delayed ambulance and denial of medical care
	26. Delayed ambulance
	27. Denial of medical care
	28. Denied ambulance
	29. Temporary denial of medical care (delayed admittance)
	30. Co-morbidity
Unknown deaths can be in any of the three (amenable, preventable or DRG) categories	31. Unknown
COVID-19 deaths can be in any of the three (amenable, preventable or DRG) categories	32. COVID-19

thus making it a preventable death or (iii) died because of a disaster risk governance failure. For example, a patient suffering from COVID-19 is being turned away from six hospitals due to a lack of available beds. Although the cause of death would be COVID-19, this is an amenable death. This is because if the patient had been admitted in time and received medical care, this death could have been avoided (see Chap. 2, Tobias & Jackson, 2001).

Various 'causes of avoidable disaster death' were emerging in the first phase of COVID-19, and most were unprecedented. Therefore, diligent efforts were made to match them with the Government of India's standard terminology to increase the data's validity. To do so, the causes of avoidable disaster death were matched as closely as possible with the categories used by the Office of the Registrar General of India for the *Causes of Death* Reports and by the National Crime Records Bureau for the *Accidental Deaths and Suicides* Report (discussed in Chap. 4). For example, the *Causes of Death Statistics 2010–2013* (ORGI, 2014) Report published by the Office of the Registrar General of India uses the terminology "motor vehicle accidents" for crashes. Similarly, the *Accidental Deaths and Suicides in India 2018* Report published by the National Crime Records Bureau (NCRB, 2019a) includes suicides and poisoning. However, some terminology does not exist in these reports, but it was found during data collection (See Table 6.2). For example, "hunger", "starvation", "exhaustion", "low hospital bed availability", "pressure of work", "stigma" and "denial of medical care" as causes of avoidable disaster death. Although "police brutality" is not mentioned in any of the reviewed Reports (Chap. 4), the statistics for "custodial crimes" are available in the *Crimes in India* Report (NCRB, 2019b). However, police brutality (*aka "lathi-charge*" or baton charge) as a term was used instead to indicate that the victims were not always in police custody.

Ascertaining the Causes of Avoidable Disaster Death

As discussed in Chap. 4, there are two methods to ascertain the causes of death. One includes using a verbal autopsy tool and a sample registration system, and the other includes medical certification of cause of death using postmortem. In both methods, a medical officer or physician ascertains the cause of death. This is the standard practice in India and other lower-middle and low-income countries (see Chap. 4, Ray-Bennett et al., 2024). This standard practice was beyond the remit of this small-scale prospective study led by me (an academic) as opposed to 'causes of death' studies (largely retrospective) led by governmental departments and health or demographic surveillance systems (for details, see Ray-Bennett et al., 2024). Furthermore, this study is based on reviewing secondary data whilst COVID-19 was still spreading along with its impact on the population. Therefore, I developed a novel strategy to ascertain the causes of death based on the emerging data (articles). This included four different actors who were ascertaining the causes of avoidable disaster deaths, and they are:

Table 6.3 Ascertaining cause of avoidable disaster death

The cause of avoidable disaster death reported was confirmed by	Number of articles (N = 252)
The author/journalist	147
The author/journalist uses statements from the police	68
The author/journalist uses a statement from a medical officer	34
Content analyst	3

Table 6.4 Number of deaths by indicator

Total number of accepted articles—252	Disaster Risk Governance—325
	Preventable—117
	Amenable—78

(i) Author/journalist (only) of the article ascertaining the cause of death
(ii) Author/journalist using a statement of a medical officer
(iii) Author/journalist using a statement of a police officer
(iv) Content analyst when there is no information supplied by (i)–(iii) (see Table 6.3).

Of the 252 articles, the author mentioned the causes of avoidable disaster death in 147 articles; in 34 articles, the author used or quoted a statement of a medical officer; for another 68 articles, the author used statements of the police. In only three articles, the author did not explain the causes of avoidable disaster death or use any statements from a medical officer or a police officer. For these articles, the content analyst/a physician and I jointly decided the causes of avoidable disaster death in alignment with the 32 broad categories introduced above (see Table 6.2). The findings also suggest that most articles do not report statements of the police or medical officers to ascertain the causes of avoidable disaster death. Compared to a medical officer's statement, a statement from a police officer was often used to ascertain the causes of death by the author.

Findings: Number of Avoidable Disaster Deaths

The content of 252 selective articles was reviewed twice to ascertain the accuracy of this data. After a double review,[2] a total of 520 avoidable disaster deaths were reported. Diligent efforts were made to count each death once. Of these 520 avoidable disaster deaths, 78 deaths were classified as amenable, 117 as preventable, and 325 as disaster risk governance-related deaths (see Table 6.4). Therefore, most avoidable disaster deaths were caused by the failure of disaster risk governance.

[2] One of the reviewers was a physician.

Findings: Circumstances of Avoidable Disaster Deaths

Chapter 3 introduced a ten-matrix of 'circumstances of death' (see Table 6.5). Based on the data, an additional indicator was added: the deceased's vulnerability. This matrix was used to review 252 accepted articles and the findings of this is provided below using each heading.

Mechanism (direct vs. indirect death): This study was designed to capture indirect deaths only. However, about 16–19 articles were considered to include direct deaths from COVID-19.

The location of deaths (indoor vs. outdoor): 146 deaths were reported to have taken place indoors, 85 outdoors and for the remainder, it was unknown whether they were indoor or outdoor because the information was not provided. The findings suggest that most deaths took place indoors, and the death was implemented mainly through 'suicide'.

The location of deaths (home, hospital or *en route*): About 68 articles mentioned deaths in public spaces or on public transport; 64 mentioned deaths at home; 18 articles mentioned deaths en route to home, a medical facility or an unknown location; 16 in unknown locations; and one article mentioned death in a police station. In summary, most deaths took place outside of medical facilities or in a non-hospital set-up.

Activity at the time of death: Detailed information for this indicator was challenging to decipher from the articles. However, a few examples of how death was implemented were provided. For instance, 68 by hanging; 44 were conducting an unknown activity; 18 jumped to commit suicide; 16 while pursuing hospital admission; nine consumed poison; two lighted themselves on fire; and one each died in police custody, struck their head off a grill, cut their wrists, and drowning (among others).

Timing of death: 68 deaths were reported to have taken place during the day, 18 deaths in the evening, and 59 at night. The timing of death in 107 articles was not mentioned. Leaving aside the 107 articles, it is suggested that most deaths took place in the evening to night hours.

Table 6.5 Matrix of circumstances of death

1. Mechanism (direct vs. indirect deaths)	2. Location of deaths (indoor vs. outdoor)	3. Location of deaths (home or hospital or enroute)
4. Activity at the time of death	5. The timing of death (day vs. night)	6. Whether the deceased person was aware of the disaster (lack of judgement)
7. Whether s/he attempted to take safety measures	8. Gender of the deceased (male/female/other)	9. Age (0–19 years, 20–51 years, and > 51 years)
10. Vulnerability of the deceased person (disability, children etc.)	11. Cause of death	

Whether the deceased person was aware of the disaster: 207 articles mentioned that the deceased person was aware of COVID-19; it was unknown for 26 articles, and 19 articles mentioned that the victims were not aware. Some cases reported below (also introduced above) show that many had little choice but to take the arduous journey from cities to their villages after the first lockdown was announced.

Whether s/he attempted to take safety measures: The safety measures, whether taken or not, were unknown in 136 articles. Eighty-seven articles indicated that the deceased person attempted to take safety measures, and 29 did not.

Gender of the deceased: Of the 520 avoidable disaster deaths, 230 were identified as male, 177 were unknown, 112 were female, and one was transgender. With this limited evidence and leaving aside the deaths of unknown gender identities of 177 people, the burden of avoidable disaster deaths was carried largely by men compared to women.

The age range of the deceased: 33 deaths were reported for the age range 0–19 years (six were below 5 years of age); 141 deaths in the age range of 20–51 years; and 64 deaths in the age range of 52 and above. Avoidable disaster deaths, therefore, occurred in the range of a few days of age up to 92 years. Another 282 deaths are recorded under the "unknown" category. For this death category, the ages and genders of the deceased are unknown. This is because the authors of these articles did not report this information. Thus, the age and gender data do not represent all the deaths.

Vulnerability of the deceased person: Not all the articles identified the identifiable attributes of the deceased. However, based on the available information, we found *ten* different categories. This included transgender (1), prisoner (1), pre-existing medical conditions (1), disabled persons (2), children (4), pregnant women (4), elderly (8), frontline workers (10), and migrant workers (46). With this limited evidence, migrant workers were the most vulnerable and carried the burden of avoidable disaster deaths in the first phase.

Cause of death: Using the statements of four different actors, we identified the 32 broad categories of causes of death (see Tables 6.4 and 6.3, respectively). Of these 32 broad categories, six categories match the death classification of the National Crime Records Bureau, and they are known as "Other causes of accidental deaths". This includes accidental electrocution, road accidents, rail accidents, suicide, poisoning, and suffocation. Two causes of death, especially medical negligence, and police brutality (both under the ambit of governance-related deaths) can be classified as 'crimes' according to the Indian Penal Crimes of Medical Negligence, and 'injuries sustained due to physical assault by police' (Category a) under the Indian Penal Codes and Special and Local Laws, respectively (NCRB, 2019b).

Limitations of Data

The data sourced from newspapers and magazines, however, are not without limitations despite having a review team consisting of a public health expert-cum-physician. The number of avoidable disaster deaths (520) does not represent the whole of India, which consists of 28 States and eight Union Territories. The Indian States publish hundreds of non-English newspapers and magazines daily. This study selected only a handful of English and reputable newspapers and magazines. Furthermore, it is essential to acknowledge the potential biases and errors of the selected sources while deriving meaningful conclusions. Although newspapers and magazines are invaluable resources for historical and current affairs, they publish news under tight deadlines. As such, errors in reporting, fact-checking or misinterpretation of information can occur, impacting the reliability of the content (Educational Wave, 2024). This was evident in many newspaper articles where vital information about the cause and circumstances surrounding deaths was incomplete. Furthermore, journalists/authors or newspapers reporting deaths may also have political leanings, affecting how news is presented and potentially leading to one-sided viewpoints (Educational Wave, 2024). Understanding these biases and limitations is crucial to uphold the integrity of the findings. Despite these limitations, this study has a few merits.

The primary purpose of this book is to introduce the avoidable disaster deaths approach to make sense of deaths that occurred in the first phase of COVID-19 so that they can be avoided in the next disaster. Furthermore, the aim is to test the analytical advantage of this approach and extrapolate specific causes of death and their circumstances to improve disaster response and preparedness systems at local and national levels. The data collection procedures and the number of deaths captured for each component indicate that the avoidable disaster deaths approach is practical, and passionate researchers and individuals can initiate small-scale projects to capture death data. Still further, this book aims to document the poignant stories of avoidable disaster deaths. People's lives were cut short by the impact of COVID-19 and the economic lockdown. Many dreams and aspirations remained unfulfilled for the deceased and their loved ones. I provide a summary of 29 cases of avoidable disaster deaths below. Of the 29 cases, 10 are risk governance-related deaths, nine are preventable deaths, six are amenable deaths, two are unknown deaths, and one is COVID-19 deaths. In the section above, I have provided the results related to the matrix of circumstances. I hope that by reading these 29 cases, readers will be able to appreciate the depth and breadth of suffering individuals had to endure and succumb to and what is needed to usher change for the next disaster. If any of the readers have lost their loved ones due to the COVID-19 pandemic, I recommend that they skip to the concluding section. This is because reading these cases can be emotional and disturbing and can bring back memories of those who are no longer with us.

Twenty-Nine Cases of Avoidable Disaster Deaths

Disaster Risk Governance

The total number of entries is 111, of which 10 (9%) cases are presented.

Case 1: One District Hospital in West Bengal has been accused of negligence as a pregnant woman died after delivering a child (The New Indian Express, 2020a). According to the deceased husband, he admitted his wife to the hospital about four to five days ago. But she received no treatment since then. She died in a wheelchair en route to her labour room. According to the deceased husband, his wife experienced a lot of pain, and she died "because of a heart attack after experiencing tremendous pain. It is the negligence of the doctor [...]" (The New Indian Express, 2020a).

Case 2: A 21-year-old migrant worker from Ratnagiri was working in Nalasopara in Maharastra to fend for her family (Suryawanshi, 2020). During the lockdown, she and her family walked on foot for almost 230 km in scorching heat to reach their home. Unfortunately, she collapsed and died on the road due to extreme heat and exhaustion (Suryawanshi, 2020).

Case 3: A family of four (husband, wife and two children were killed by a speeding vehicle from the opposite side of the road (Scroll Staff, 2020a). The husband was a migrant worker and the family was en route to their village by cycle. The couple was taken to a local hospital but succumbed to their injuries. However, their children—a three-year-old son and a four-year-old daughter—survived the crash and were admitted to a hospital in Lucknow (Scroll Staff, 2020a).

Case 4: In Madhya Pradesh's Guna district, a truck carrying migrant workers collided with a bus (Scroll Staff, 2020b). This collision killed eight migrant workers and injured 54. The migrant workers were en route to their hometowns. The deaths were confirmed by the Additional Superintendent of Police (Scroll Staff, 2020b).

Case 5: A 12-year-old girl died after walking for three days with other labourers. She was returning home during the lockdown (Thakur & Painkra, 2020). The deceased belonged to the Muria community of Adivasis of Chhattisgarh. She was a chilly picker and earned either Rs. 200 (~ USD 2.37) a day or sacks of chillis as pre-arranged payments. She took the forest route because the main road was closed. They walked 100 kms in three days, hungry and exhausted. On April 18, around 9 a.m., 60 kms from home, the girl died. She complained of stomach pain and headache. Also, the night before, she fell and broke a bone. The authors mentioned that the medical report was not available to confirm the cause of death (Thakur & Painkra, 2020).

Case 6: A woman of 35-year-old died on board the Ahmedabad-Katihar Shramik Special train (Thakur, 2020). The body was kept on the platform till an ambulance arrived to take the deceased to her home in Katihar district. The deceased child was seen pulling up the sheet of cloth covering his mother's body to wake her up. The deceased's brother-in-law told the media that she died due to hunger and severe heat. However, the Muzaffarpur district administration, as well as the Railways, claimed that she died because of a pre-existing ailment (Thakur, 2020).

Case 7: A 32-year-old man in West Bengal's Howrah district died hours after he was beaten by police for stepping out of his house to buy milk during the nationwide lockdown (The Wire Staff, 2020). According to his family, the man died of the injuries sustained during the police's assault on him. The victim was rushed to a local hospital, where he was declared dead. The police have denied the claim that the man died due to the *lathi-charge* (baton-charge) and instead blamed a cardiac arrest, saying the deceased was already suffering from heart ailments (The Wire Staff, 2020).

Case 8: A 27-year-old youth committed suicide at a Covid Care Centre in Puliyangudi. Sources said the youth took the extreme step fearing the social stigma associated with the disease (Tamil Nadu, 2020).

Case 9: A 17-year-old migrant girl from Madhya Pradesh ended her life after she and her four siblings were left starving for two days during the lockdown (Nation, 2020a). One Police Inspector said an inquiry revealed that this migrant family from Madhya Pradesh had no food for the last two days. The neighbours gave them food, but the girl was ashamed of her plight and was humiliated to beg every day. According to the Police Constable, when her father was away in search of work, the deceased asked her siblings to go out. She hung herself in their tiny room (Nation, 2020a).

Case 10: According to a police report, a train killed 14 migrant workers who had fallen asleep on the track while they were heading back home after losing their jobs amid the coronavirus lockdown (CBC News Staff, 2020). Police said the labourers worked for a steel company and were returning home to Madhya Pradesh (CBC News Staff, 2020).

Preventable Deaths

The total number of entries is 90, of which nine stories (10%) are presented here.

Case 1: A young, meritorious student of class X took her life by setting herself ablaze, according to her parents (Sambad English Bureau, 2020a). The girl could not attend the online classes during the lockdown, which made her very gloomy. The autopsy would be conducted later at the State-run Medical College hospital and would be handed to the family to conduct the last rites. The girl's father was a casual labourer and on account of the lockdown, had no work. Hence, he lacked money to repair their TV or buy a smartphone for the online classes (Sambad English Bureau, 2020a).

Case 2: An 82-year-old woman was found dead inside a toilet in the same hospital where she was admitted according to the senior police inspector and her relatives (Sambad English Bureau, 2020b). The woman had tested Covid-19 positive on May 27 and had been admitted to another hospital before she was shifted to the current hospital. Her relatives informed the police that she was missing on June 2. In a video message, the deceased grandson appealed to the Chief Minister to probe into the incident and punish those found negligent or guilty (Sambad English Bureau, 2020b).

Case 3: A 20-year-old migrant worker allegedly committed suicide at a quarantine centre in Chhattisgarh's Gariaband district, where he was kept after he returned from Telangana, police said (Nation, 2020b). The deceased was found hanging in a room at the quarantine centre in his native village. According to the Additional Superintendent of Police, the migrant worker had returned from Hyderabad a week before and was kept at the quarantine facility for 14 days as a precautionary measure. "He did not have any coronavirus symptoms. However, his sample has been sent for COVID-19 test as per protocol" according to the Additional Superintendent of Police (Nation, 2020b).

Case 4: According to a Circle Officer, a 19-year-old migrant labourer who had returned to Uttar Pradesh from Gujarat amid the lockdown allegedly committed suicide by hanging himself from the ceiling fan in his room. The body was handed over to the deceased family members after postmortem, according to this Officer. The family members could not tell the motive behind the suicide (Nation, 2020c).

Case 5: A 35-year-old migrant worker, who had returned from Mumbai recently, committed suicide by hanging himself from the ceiling at a COVID-19 hospital in the morning hours after testing positive for coronavirus on the previous day. Doctors said that the man killed himself after the four persons he was sharing the facility with were discharged after their reports came back negative. The deceased was asked to remain in isolation for a few more days till the final report came for the test samples sent to Ranchi (Nation, 2020d).

Case 6: A 62-year-old woman allegedly committed suicide by jumping into the Krishna River after her relatives abandoned her, suspecting that she could have contracted COVID-19 (Srinivas, 2020). The deceased lost her son due to COVID-19, and her daughter-in-law was shifted to a quarantine centre as she tested positive. She went to her brother's house seeking help but was reportedly not allowed to enter the house. According to a local resident, "She was seen speaking in a disturbing mood on Sunday and cried as nobody was performing her son's funeral. Later, she went missing." The local Member of the Legislative Assembly visited and expressed concerns over the inhuman attitude of the neighbours and relatives (Srinivas, 2020).

Case 7: Just two days after returning to their home district in Odisha during the COVID -19 lockdown, a 33-year-old woman hung herself to death out of grief in a state-run hospital after she lost her husband under treatment. According to a medical officer-in-charge, the deceased attempted to kill herself by slitting her hand after her husband passed away, but the ward staff managed to prevent her from committing the act. However, they failed to bring her to a normal state. The woman then closed herself inside the room and hung herself from a ceiling fan in front of her husband's body, according to the medical officer in-charge (Sambad English Bureau, 2020c).

Case 8: A journalist succumbed to his injuries after he committed suicide by jumping off the fourth floor of a hospital in Delhi, where he was admitted after testing positive for Covid-19 (Mani, 2020). He was critically injured and was admitted to the ICU. After testing positive for coronavirus, he was sending depressing messages to his friends and colleagues, sources said. According to his friend, "Due to lockdown and Covid, he feared he would lose his job. He was going through a financial crisis also. He had complained that his boss was mentally torturing him." The deceased is survived by his wife and two daughters (Mani, 2020).

Case 9: According to a district superintendent of police in Andhra Pradesh, at least ten alleged alcohol addicts (three of whom were beggars and two were rickshaw pullers) died after consuming sanitiser as an alternative to liquor (PTI, 2020). All the deceased were alcohol addicts, and its non-availability due to the extended COVID-19 lockdown forced them to go for the alcohol-based sanitiser, according to the District Superintendent of Police. According to this police personnel, they are investigating whether the sanitiser had any other toxic content. To assess this, they have sent the stocks for chemical analysis (PTI, 2020).

Amenable Deaths

The total number of entries is 60, of which six cases (10%) are presented here.

Case 1: The body of a newborn baby was recovered from the bag of his parents at a transit centre for migrant workers in Odisha. The baby's parents worked at a brick kiln in Uttar Pradesh. After the lockdown was enforced, the couple registered to return to their village by train. The baby was born in a hospital in Uttar Pradesh a day before their departure. The couple boarded the train the next day with their newly-born baby boy. Late at night, they noticed that the baby had stopped crying and found he had probably died of suffocation in the overcrowded train. Apprehending that other passengers might force them out of the train for carrying the dead baby, the couple put the deceased in one of their bags. After reaching their destination the couple were shifted to a transit centre. In the evening, the body was found by someone at the centre and police were informed. Police seized the body and interrogated the couple. After the postmortem, the couple were sent to their village along with the body. Further investigation will follow once the postmortem report is handed over to the police (Odisha, 2020).

Case 2: In the Ganjam district of Odisha, a critically ill man died despite the arrival of the 108 ambulance (PNN, 2020). His family members watched him die helplessly after a 108-ambulance refused to carry the person to the hospital, suspecting him of suffering from COVID-19. According to the deceased son: "The 108 ambulance employees said the 104 ambulances will arrive at our house and carry my father to the hospital. But the 104-ambulance reached here late, and in the meantime, my father was dead." After the media highlighted the issue, The Ganjam District Collector issued an order stating that all 108/104 ambulances will be used for shifting suspected COVID-19 cases to the appropriate health facility during emergencies in adhering to IPC guidelines of COVID-19 (PNN, 2020).

Case 3: A youth who was kept at a quarantine centre in the Balasore district of Odisha died at a Community Health Centre following illness as an ambulance allegedly failed to turn up (Sambad English Bureau, 2020d). The deceased and his brother were migrant workers and had returned from Mumbai during the lockdown. Both were placed under quarantine. The deceased complained of sickness the night before, after which his brother contacted 108 Ambulance Service. It is alleged that the ambulance service staff refused to carry the patient from the quarantine centre to the hospital. With no other option left, the deceased brother took him to a Community Health Centre on a motorcycle. However, the person died while undergoing treatment at the Community Health Centre. The deceased had heart-related issues. According to a medical officer of the Community Health Centre, the cause of death can be ascertained after post-mortem (Sambad English Bureau, 2020d).

Case 4: A few hours after arrival from Andhra Pradesh, a 6-year-old boy died at a quarantine centre in the Nuapada district of Odisha (Sambad English Bureau, 2020e). According to reports, the child along with his parents, who are migrant labourers, were at a quarantine centre. The child was suffering from fever and loose motion. According to an on-duty official of the Temporary Medical camp "As we were told [by ASHA (Accredited Social Health Activist) worker and the temporary medical camp (TMC)] that the condition of the child patient was not serious, we did not make immediate arrangements for medical care." Expressing anguish over the death of the child, the parents of the deceased alleged negligence by the local authorities for which they lost their son. "Neither the ailing child was given medical care nor were we served food on time," said the father of the deceased (Sambad English Bureau, 2020e).

Case 5: A 30-year-old pregnant woman requiring urgent medical attention in the eighth month of her pregnancy died in Ghaziabad (Nation, 2020e). The deceased and her husband travelled from one hospital to another for 13 h. Eight hospitals, including the government ones, had allegedly turned them away. The deceased brother-in-law said: "We want justice [...]. A penalty on hospitals is not justice. [...]. A patient should get admitted to a hospital and get treatment. [...] Life and death are in the hands of the Almighty, but at least they can try. [...] Nobody should die because of lack of treatment or health care" (Nation, 2020e).

Case 6: A 27-year-old migrant worker died in a Shramik Special train as he returned to Malda from Kerala (Statesman News Service, 2020). The deceased and his cousin were returning home during the countrywide lockdown. Police said they had sent the body for a post-mortem and that it was yet unclear if the worker had contracted Covid-19. According to the deceased cousin, he complained of abdominal pain for two days, but the train did not stop anywhere for his treatment. Furthermore, they could not manage pure drinking water for him. "This morning, he did not respond to our wake-up calls and we came to know only later that he had already died," according to the deceased cousin. The deceased is survived by his wife and two sons, six and four years of age. The District President of the Trinamul Congress party blamed the railways for the death for not having medical support for those who fall ill on the train (Statesman News Service, 2020).

Unknown Deaths

The total number of entries is 23, of which two cases (~ 9%) are presented here. Unknown death is understood here as death due to amenable, preventable, and governance-related.

Case 1: A doctor working with the health department in Karnataka hung himself to death at his home allegedly due to intense work pressure owing to the Covid-19 pandemic (Pinto, 2020). The deceased colleagues and relatives claimed there was a lot of administrative pressure from higher-ups and political bosses (Pinto, 2020).

Case 2: A 50-year-old unemployed man, due to the lockdown, committed suicide in Uttar Pradesh (Pandey & Verma, 2020). He left behind a suicide note that blames India's extended lockdown over the coronavirus pandemic for his inability to care for his four children, his wife and his unwell mother. The body, along with a suicide note, was found on a rail track. The deceased was unemployed since the lockdown and had reportedly run out of cash in the last few days. In the suicide note, the man said that while there was wheat and rice in his house, thanks to the government ration shop, it was not enough. He said he did not have money to buy other essentials like sugar, salt, and milk (Pandey & Verma, 2020).

COVID-19 Deaths

The total number of entries is 16, of which one case (~ 7%) is presented here. COVID-19 death is understood here as death due to amenable, preventable and governance-related.

Case 1: A 19-year-old girl travelling with her mother on a UP roadways bus from Delhi to Shikohabad was allegedly thrown out by the driver and conductor on Yamuna Expressway after passengers thought she exhibited "symptoms of COVID-19" (Jaiswal, 2020). The girl died on the road 30 min later. According to the deceased brother, "the driver and conductor threw a blanket over my sister as they didn't want to touch her and dragged her out of her seat. She was not feeling well as it was very hot that day. My mother pleaded with them, but they did not listen, and no one helped them. [...] My sister was in good health and didn't have any disease. She had kidney stones earlier but had recovered. How can a healthy individual drop dead just like that?" An ambulance took the body for a post-mortem (Jaiswal, 2020).

Conclusion

Professor Sridhar (2021, p. 220) posed a mooting question in her book *Preventable: How a Pandemic Changed the World and How to Stop the Next One*: "**Was the cure (lockdown) worse than the disease?**" This is a fitting question in the context of this study. If not lockdown, what could have been the alternative for India—home to almost the world's largest and poorest people. The Oxford COVID-19 Government Response Tracker, which 'measures the strictness of the lockdown imposed', gave the highest score (100) to India's Lockdown 1.0 (University of Oxford, 2020; quoted in Thayer et al., 2021). This Lockdown 1.0 was one of human history's most considerable concerted efforts at social distancing (Umachandran, 2020).[3] This Lockdown also triggered a substantial economic, humanitarian, and psychosocial impact, evident in the cases above. Above all, this impact created enabling circumstances for avoidable disaster deaths. To conclude this chapter, a brief discussion on the impact of the lockdown is presented below.

Economic impact: According to Sidhu et al. (2020), the pandemic had a broader impact than the recession of 2008 in India. The worst-hit industries included tourism, hospitality, transport, oil, entertainment, real estate, advertising, aviation and construction. Indian real gross domestic product contracted by – 23.9% in the first quarter of 2020–21, compared to a 5.2% growth in the previous year's first quarter (Government of India, 2020). This contraction is the worst recorded in history (Sridhar, 2021). Private consumption fell by 26.7%, and investment demand fell by 47.1%. Further, a fall of – 50.3% was seen in construction, – 39.3% in manufacturing, – 23.4% in mining and – 47% in the tourism and hotel industry (Sridhar, 2021). These sectors rely on economic migrants. However, agriculture saw growth at a rate of 3.4% (Government of India, 2020).

The economic fallout from COVID-19 has been widespread. Thus, all individuals and households felt the impacts of COVID-19 regardless of whether or not one contracted the virus. That said, the socio-economically disadvantaged sections of society, especially migrants, were the hardest hit, with no economic relief package in time. In many instances (presented above), some succumbed to death by starvation, others resorted to begging to survive, and still others, unfortunately, died whilst taking an arduous journey to return home.

Humanitarian impact: The first two lockdowns triggered a "humanitarian disaster of unprecedented proportions" (Sengupta & Jha, 2020, p. 152). According to Jose (2021) this is partly because of the government's own doing "specifically through the massive disruption caused by its ill-conceived and worse-implemented—national lockdown in March 2020". About 90% of the Indian labour force is employed in the unorganised sector, and millions work in urban areas. It is estimated that about 100 million migrants, predominantly daily wage earners with no job security, work across India's large, medium, and small-sized cities (Economic Times, 2020).

[3] The density of the population of India is 464 persons per square metre, which makes the practice of social distancing extremely challenging. Social distancing is rather a privilege and "only possible if you have a large house" (Pandey, 2020).

Migration in India mainly occurs from rural to urban areas, and the migrants are primarily poor (Pandey, 2020). They migrate in search of livelihoods or for diversification of income. They work as domestic helpers, drivers, gardeners or cooks, daily wage earners on construction and manufacturing sites, building malls, flyovers, and homes, or as street vendors—among other jobs (Pandey, 2020). Sengupta and Jha (2020, p. 166) call them 'the drivers' of the 'engine of globalisation'. According to the Economic Times (2020), these workers, the 'engine of globalisation', were the worst affected by the lockdown restrictions between March and early June 2020. Their plight of walking hundreds of miles under the sweltering heat of March with their mere belongings or children or both on their backs attracted global attention.

Millions of migrant workers covered hundreds of kilometres on foot, on cycles, and on the back of lorries or tuk tuks, hungry, thirsty and with no or very little money. There was no transportation available for them when Lockdown 1.0 was announced. En route, many met persecution and violence at the hands of the police who were implementing the Lockdown 1.0 orders (Sengupta & Jha, 2020). Despite the lockdown, road and motor vehicle accidents increased, disproportionately affecting migrant workers. By the end of May (two months after the implementation of the lockdown), almost 200 migrant workers succumbed to injuries or deaths from road accidents (Banerji, 2020). Other significant causes of death of migrant workers were hunger and starvation as well as suicide due to reasons such as economic distress (Elsa, 2020). This trend is consistent with the findings of this study.

The Ministry of Home Affairs (2020) instructed the Union and State Governments to ensure adequate arrangements of food and shelter for migrant workers, as well as operating Special Trains (aka Shramik Train) for those who were stranded, including migrant workers (PIB, 2020). However, these measures were "inadequate and insufficient due to the vast majority of internal migrant workers not qualifying for relief packages, and the lack of coordination among state governments for the transportation of internal migrants" (OHCHR, 2020). For some, they were too late.

The full extent of the impact of the sudden lockdown on the working population that form India's majority is incalculable (Jose, 2021). However, the Centre for Monitoring Indian Economy estimates that the abrupt shutdown led to nearly 140 million jobless people (Jose, 2021). It was also estimated that India (amongst the other ten countries where extreme poverty is high) is likely to see a spike in extreme poverty with an additional 85 million people and a revised growth rate of − 11%, one of the deepest recessions in the world (Kharas, 2020). This will alter India's poverty trajectory, which has been trending downwards (Kharas, 2020).

Psycho-social impact: Unemployment, food insecurity, and the fear of contracting and spreading the virus during the lockdown have had huge psycho-social impacts on individuals and households across the country. Isolation led to significant mental health risks, including but not limited to anger, sleeplessness, and fear in the initial few days of isolation, and depression and post-traumatic stress disorder in the long run. Stressful experiences such as learning about the diagnosis of COVID-19, fear of infecting others, symptoms of the illness, hospitalisation, especially admission to an intensive care unit, and loss of income led to the development of anxiety, depression, and post-traumatic stress disorder (Sher, 2020) and in some circumstances in-human and apathetic behaviours (see Case 1 above under the theme COVID-19 deaths).

According to a survey conducted by Grover et al., (2020), more than two-fifths of the participants experienced anxiety and depression because of the lockdown and COVID-19 in India. Nearly half of the participants reported an increased prevalence of "negative emotions such as sadness, loneliness, anxiety, frustration, and fear and apprehension" (Grover et al., 2020, p. 361). Social isolation also contributed to suicidal behaviour. Those diagnosed with COVID-19, and those who have survived it, especially those with severe symptoms, were found to be at an elevated risk of suicide across India[4] (The Times of India, 2020; The New Indian Express, 2020b). Suicide rates rose "exponentially" in regions such as Himachal Pradesh, where 466 persons committed suicide between February and August 2020 (Sharma, 2020a). As of July 2020, the lockdown was the leading cause of over 300 non-COVID-19 deaths according to Sharma (2020b). This trend is consistent with this study's findings. Suicide under the ambit of preventable deaths carried the largest burden of avoidable disaster deaths.

Considering the economic, humanitarian, and psychosocial impact triggered by COVID-19 and the lockdown, as well as its continued effects, **"was the cure (lockdown) worse than the disease?"** The answer is YES. This is because it led to non-COVID-19-related excess deaths. This cure (lockdown), however, would have been effective if it was done in consultation with the public (Sen, 2020) and by developing calculative phases that support the most vulnerable groups through economic packages, preventable and amenable public health packages, health systems and by learning lessons from neighbours including South Korea, Senegal (discussed in next chapter) [and even Kerala] who flattened the curves without lockdown.

At the height of the first wave of COVID-19, Noble Laurette, Professor Amartya Sen, wrote a piece on 'Listening as Governance', which is succinct in this context. Sen wrote (2020): "Tackling a social calamity is not like fighting a war which works best when a leader can use top-down power to order everyone to do". To overcome a social calamity, listening is a necessity, according to Sen (2020). This applies to the calamity caused by the pandemic and by understanding that some groups of people are more vulnerable than others, and this is largely due to an individual's level of exposure to the risk or other characteristics such as class, caste, and gender—among other vulnerabilities. The different types of hazards that different groups suffer need addressing, which can be aided by participatory democracy through free and public discussion, according to Sen (2020).

---

[4] Evidence suggests that deaths by suicide increased in the United States during the 1918–19 influenza pandemic and among older people in Hong Kong during the 2003 severe acute respiratory syndrome (SARS) epidemic (Gunnell et al., 2020).

References

Banerji, A. (2020). Nearly 200 migrant workers killed on India's roads during coronavirus lockdown. *Reuters*, 2 June. Available at: https://www.reuters.com/article/us-health-coronavirus-india-migrants/nearly-200-migrant-workers-killed-on-indias-roads-during-coronavirus-lockdown-idUSKBN2392LG. Accessed on September 8, 2020.

CBC News Staff. (2020). Train kills 14 labourers laid off in coronavirus lockdown in India. *CBC News*, 8 May. Available at: https://www.cbc.ca/news/world/india-train-crash-1.5560926. Accessed on May 8, 2020.

Economic Times. (2020). What a coronavirus lockdown looks likes, and what you can do & what you can't. *The Economic Times*, 25 March. Available at: https://economictimes.indiatimes.com/news/politics-and-nation/coronavirus-outbreak-what-a-lockdown-will-look-like-for-you/articleshow/74760719.cms. Accessed on September 5, 2020.

Education Wave. (2024). *Pros and cons of using newspapers for research*. Available at: https://www.educationalwave.com/pros-and-cons-of-using-newspapers-for-research/. Accessed on October 12, 2024.

Elsa, E. (2020). #MigrantLivesMatter: India's migrant workers are dying due to hunger, suicides and accidents, in coronavirus lockdown. *Gulf News*, 19 May. Available at: https://gulfnews.com/world/asia/india/migrantlivesmatter-indias-migrant-workers-are-dying-due-to-hunger-suicides-and-accidents-in-coronavirus-lockdown-1.1589896799187. Accessed on September 8, 2020.

Government of India. (2020). *Lok Sabha unstarred question No 174, answered on 14.09.2020: Migrant workers returning to hometown*. Government of India. Available at: http://164.100.47.194/Loksabha/Questions/Qresult15.aspx?qref=16409&lsno=17. Accessed on September 16, 2020.

Grover, S., Sahoo, S., Mehra, A., Avasthi, A., Tripathi, A., Subramanyan, A., Pattojoshi, A., Rao, G., Saha, G., Mishra, K., Chakraborty, K., Rao, N., Vaishnav, M., Singh, O. P., Dalal, P. K., Chadda, R., Gupta, R., Gautam, S., Sarkar, S., & Sathyanarayana Rao, T. S. (2020). Psychological impact of COVID-19 lockdown: An online survey from India. *Indian Journal of Psychiatry, 62*(4), 354–354. https://doi.org/10.4103/psychiatry.indianjpsychiatry_427_20

Gunnell, D., Appleby, L., Arensman, E., Hawton, K., John, A., & Kapur, N. (2020). Suicide risk and prevention during the COVID-19 pandemic. *The Lancet, 7*(6), 468–471.

Jaiswal, A. (2020). Delhi girl thrown out of bus on Agra Expressway over Covid suspicion, dies. *The Times of India*, 10 July. Available at: https://timesofindia.indiatimes.com/city/agra/delhi-girl-thrown-out-of-up-bus-over-covid-suspicion-dies/articleshow/76881420.cms. Accessed on July 20, 2020.

Jose, S. (2021). The Covid lockdowns and the catastrophe of hunger in India—Part one. Countercurrents.org. *Countercurrents*, 21 August. Available at: https://countercurrents.org/2021/08/covid-and-the-catastrophe-of-hunger-in-india-part-one/. Accessed on August 21, 2022.

Kharas, H. (2020). *The impact of COVID-19 on global extreme poverty*. Brookings. Available at: https://www.brookings.edu/blog/future-development/2020/10/21/the-impact-of-covid-19-on-global-extreme-poverty/. Accessed on September 17, 2022.

Mani, G. (2020). Covid-19 positive journalist dies by suicide at AIIMS Delhi, claimed injections administered. *The New Indian Express*, 6 July. Available at: https://www.newindianexpress.com/nation/2020/jul/06/covid-19-positive-journalist-dies-by-suicide-at-aiims-delhi-claimed-injections-administered-2166145.html. Accessed on July 6, 2020.

Ministry of Home Affairs. (2020). *Annexure to Ministry of Home Affairs order no. 40-3/2020-D dated 24.03.2020*. Ministry of Home Affairs. Available at: https://www.mha.gov.in/sites/default/files/MHAorder%20copy_0.pdf. Accessed on September 5, 2020.

Nation. (2020a). Left starving for two days, 17-year-old migrant girl commits suicide in Maharashtra. *The New Indian Express*, 12 May. Available at: https://www.newindianexpress.com/nation/2020/may/12/left-starving-for-two-days-17-year-old-migrant-girl-commits-suicide-in-maharashtra-2142503.html. Accessed on May 12, 2024.

Nation. (2020b) Telangana-returnee migrant hangs self in Chhattisgarh's coronavirus quarantine centre. *The New India Express*, 7 June. Available at: https://www.newindianexpress.com/nation/2020/jun/07/telangana-returnee-migrant-hangs-self-in-chhattisgarhs-coronavirus-quarantine-centre-2153504.html. Accessed on June 8, 2020.

Nation. (2020c). 19-year-old migrant labourer who returned from Gujarat commits suicide in UP. *The New India Express*, 9 June. Available at: https://www.newindianexpress.com/nation/2020/jun/09/19-year-old-migrant-labourer-who-returned-from-gujarat-commits-suicide-in-up-2154202.html. Accessed on June 10, 2020.

Nation. (2020d). Hours after testing positive, Jharkhand migrant ends life in COVID-19 hospital. *The New India Express*, 9 June. Available at: https://www.newindianexpress.com/nation/2020/jun/09/hours-after-testing-positive-jharkhand-migrant-ends-life-in-covid-19-hospital-2154359.html. Accessed on June 9, 2020.

Nation. (2020e). Ensure no patient is denied admission: Kin of Noida woman who die shuttling between hospitals. *The New Indian Express*, 8 June. Available at: https://www.newindianexpress.com/nation/2020/jun/08/ensure-no-patient-is-denied-admission-kin-of-noida-woman-who-died-shuttling-between-hospitals-2153779.html. Accessed on June 8, 2020.

NCRB (National Crime Records Bureau). (2019a). *Accidental deaths and suicides in India*. Available at: https://ncrb.gov.in/sites/default/files/ADSI-2018-FULL-REPORT-2018.pdf. Accessed on June 15, 2020.

NCRB (National Crime Records Bureau). (2019b). *Crime in India-2018*. Available at: https://ncrb.gov.in/sites/default/files/ADSI-2018-FULL-REPORT-2018.pdf%20. https://ncrb.gov.in/en/crime-india-2018. Accessed on June 15, 2020.

Odisha. (2020). Body of new-born boy found in parents' bag at Odisha's Nuapada district. *The New Indian Express*, 19 June. Available at: https://www.newindianexpress.com/states/odisha/2020/jun/19/body-of-newborn-boy-found-in-parents-bag-at-odishas-nuapada-district-2158559.html. Accessed June 20.

OHCHR (United Nations Human Rights Office of the High Commissioner). (2020). *COVID-19: Urgent help for India's forgotten migrant workers must follow Supreme Court ruling, say UN experts*. OHCHR. Available at: https://www.ohchr.org/EN/NewsEvents/Pages/DisplayNews.aspx?NewsID=25926. Accessed on September 8, 2020.

ORGI (Office of the Registrar General of India). (2014) *Special registration system: Cause of death statistics 2010–2013*. Available at: file:///C:/Users/nibed/Downloads/SRS_COD-STATISTICS_2010–2013.pdf. Accessed on October 22, 2024.

Pandey, A., & Verma, S. (2020). Unable to care for family, UP man commits suicide, blames lockdown. *NDTV*, 30 May. Available at: https://www.ndtv.com/india-news/coronavirus-lockdown-unable-to-provide-food-for-family-uttar-pradesh-man-kills-self-in-lakhimpur-kheri-blames-lockdown-2237738. Accessed on May 30, 2020.

Pandey, V. (2020). Coronavirus lockdown: the Indian migrants dying to get home. *BBC News*, 19 May. Available at: https://www.bbc.co.uk/news/world-asia-india-52672764. Accessed on May 20, 2020.

PIB (Press Information Bureau). (2020). *Special Trains to be operated for Movement of Stranded Persons across the Country, due to Lockdown Restrictions to fight COVID-19*. Available at: https://www.mha.gov.in/sites/default/files/press.pdf. Accessed on September 08, 2020.

Pinto, N. (2020). Covid warrior commits suicide due to work pressure; CM orders probe, announces Rs 50 lakh ex-gratia. *India Today*, 22 August. Available online at: https://www.indiatoday.in/india/story/covid-warrior-commits-suicide-due-to-work-pressure-cm-orders-probe-announces-rs-50-lakh-ex-gratia-1713801-2020-08-21. Accessed on August 22, 2020.

PNN. (2020). Critically ill man dies after being denied ambulance service over COVID-19 scare. *Odisha Post*, 27 July. Available at: https://www.orissapost.com/man-dies-after-being-denied-ambulance-service-over-coronavirus-scare/. Accessed on July 27, 2020.

PTI. (2020). Andhra Pradesh: 10 villagers die after consuming Sanitiser as alternative to alcohol. *The Wire*, 31 July. Available at: https://thewire.in/health/10-people-die-after-consuming-sanitiser. Accessed on July 31, 2020.

Ray-Bennett, N. S., Dissanayake, L., Ekezie, W., & Macleod, L. (2024). *Assessment of drowning data collection processes in low-and middle-income countries: A scoping review report*. Avoidable Deaths Network and Royal National Lifeboat Institution.
Sambad English Bureau. (2020a). Unable to attend online class, class X girl sets herself on fire. *Sambad English*, 2 June. Available at: https://sambadenglish.com/unable-to-attend-online-class-class-x-girl-sets-herself-on-fire/. Accessed on June 2, 2020.
Sambad English Bureau. (2020b). 'Missing' Covid patient found dead at hospital toilet, *Sambad English*, 10 June. Available at: https://sambadenglish.com/missing-covid-patient-found-dead-at-hospital-toilet/. Accessed on June 10, 2020.
Sambad English Bureau. (2020c). Odisha woman hangs self to death after husband dies in COVID-19 isolation ward. *Sambad English*, 3 July. Available at: https://ommcomnews.com/odisha-news/odisha-woman-hangs-self-in-covid-19-ward-after-husbands-death-in-jagatsinghpur. Accessed on July 3, 2020.
Sambad English Bureau. (2020d). Quarantined ill youth dies in Odisha as ambulance fails to turn up. *Sambad English*, 4 June. Available at: https://sambadenglish.com/quarantined-ill-youth-dies-in-odisha-as-ambulance-fails-to-turn-up/. Accessed on June 4, 2020.
Sambad English Bureau. (2020e). 6-YO boy dies at COVID-19 quarantine centre in Odisha. *Samd English*, 6 June. Available at: https://sambadenglish.com/6-yo-boy-dies-at-covid-19-quarantine-centre-in-odisha/. Accessed on June 6, 2020.
Scroll Staff. (2020a). Road accident kills migrant labourer and his wife. *Scroll*, 8 May. Available at: https://scroll.in/latest/961395/covid-19-migrant-worker-wife-die-in-road-crash-in-lucknow-while-trying-to-cycle-to-chhattisgarh. Accessed on May 8, 2020.
Scroll Staff. (2020b). Covid-19 lockdown: 14 migrant workers killed in two separate accidents in Madhya Pradesh and UP. *Scroll*, 14 May. Available at: https://scroll.in/latest/961880/covid-19-lockdown-14-migrant-workers-killed-in-two-separate-accidents-in-madhya-pradesh-and-up. Accessed on May 14, 2020.
Sengupta, S., & Jha, M. K. (2020). Social policy, COVID-19 and impoverished migrants: Challenges and prospects in locked down India. *The International Journal of Community and Social Development, 2*(2), 152–172. https://doi.org/10.1177/2516602620933715
Sen, A. (2009). *The idea of justice*. Allen Lane.
Sen, A. (2020). Listening as governance. *CPNN*, 21 June. Available at: https://cpnn-world.org/new/?p=20650. Accessed on June 21, 2020.
Sharma, A. (2020a). Suicide rates rise exponentially in Himachal Pradesh as economy continues to flutter due to Covid. *Outlook India*, 10 August. Available at: https://www.outlookindia.com/website/story/india-news-suicide-rates-rise-exponentially-in-himachal-pradesh-as-economy-continues-to-flutter-due-to-covid/358385. Accessed on September 11, 2020.
Sharma, S. (2020b). COVID-19: Lockdown and financial crises lead to jump in suicide rate in UP. *NewsClick*, 30 June. Available at: https://www.newsclick.in/COVID-19-Lockdwon-UP-Financial-Crises-Suicide-Rise. Accessed on September 11, 2020.
Sher, L. (2020). The impact of the COVID-19 pandemic on suicide rates. *QJM: An International Journal of Medicine, 113*(10), 707–712. https://doi.org/10.1093/qjmed/hcaa202
Sidhu, G. S., Rai, J. S., Khaira, K. S., & Kaur, S. (2020). The impact of Covid-19 pandemic on different sectors of the Indian economy: A descriptive study. *International Journal of Economics and Financial Issues, 10*(5), 113–120. Available at: https://www.econjournals.com/index.php/ijefi/article/view/10461. Accessed on September 16, 2020.
Sridhar, D. L. (2021). *Preventable: How a pandemic changed the world & how to stop the next one*. Viking.
Srinivas, R. (2020). Elderly woman abandoned by kin suspecting COVID, 'commits' suicide. *The Hindu*, 27 July. Available at: https://www.thehindu.com/news/national/andhra-pradesh/elderly-woman-abandoned-by-kin-suspecting-covid-commits-suicide/article32206575.ece. Accessed on July 27, 2020.

Statesman News Service. (2020). Malda migrant worker dies in train. *The Statesman*, 10 June. Available at: https://www.thestatesman.com/bengal/malda-migrant-worker-dies-train-1502898294.html. Accessed on June 11, 2020.

Suryawanshi, S. (2020). After over 200 km walk under scorching heat, woman dies in Konkan. *The New Indian Express*, 8 May. Available at: https://www.newindianexpress.com/nation/2020/may/08/lockdown-30-after-over-200-km-walk-under-scorching-heat-woman-dies-in-konkan-2140711.html. Accessed on May 9, 2020.

Tobias, M., & Jackson, G. (2001). Avoidable mortality in New Zealand, 1981–97. *Australian and New Zealand Journal of Public Health, 25*(1), 12–20. Available at: https://doi.org/10.1111/j.1467-842X.2001.tb00543.x. Accessed on October 29, 2020.

Tamil Nadu. (2020). Tenkasi: Covid stigma pushes one more to end life. *The New India Express*, 5 August. Available at: https://www.newindianexpress.com/states/tamil-nadu/2020/aug/05/tenkasi-covid-stigma-pushes-one-more-to-end-life-2179218.html. Accessed on August 5, 2020.

Thakur, P., & Painkra, K. (2020). Jamlo's last journey along a locked-down road. *Pari*, 14 May. Available at: https://ruralindiaonline.org/en/articles/jamlos-last-journey-along-a-locked-down-road/. Accessed on May 14, 2020.

Thakur, R. K. (2020). Video of a toddler trying to wake up dead mother at Bihar railway station sparks outrage. *The New India Express*, 27 May. Available at: https://www.newindianexpress.com/nation/2020/may/27/migrants-crisis-toddler-tries-to-wake-up-dead-mother-at-bihar-railway-station-2148674.html. Accessed on May 27, 2020.

Thayer, W. M., Hasan, Z., Sankhla, P., & Gupta, S. (2021). An interrupted time series analysis of the lockdown policies in India: A national-level analysis of COVID-19 incidence. *Health Policy and Planning, 36*(5), 620–629. https://doi.org/10.1093/heapol/czab027

The New India Express. (2020a). Asansol district hospital accused of negligence as pregnant woman dies after delivery. *The New Indian Express*, 8 June. Available online at: https://www.newindianexpress.com/nation/2020/jun/08/asansol-district-hospital-accused-of-negligence-as-pregnant-woman-dies-after-delivery-2153753.html. Accessed on June 8, 2020.

The New Indian Express. (2020b). Surat: 63-yr-old man 'kills self' after testing positive for Covid, 11 July. Available at: https://indianexpress.com/article/india/surat-63-yr-old-man-kills-self-after-testing-positive-for-covid-6500077/. Accessed on September 11, 2020.

The Times of India. (2020).'Andhra couple commits suicide after recovering from COVID-19, 2 August. Available at: https://timesofindia.indiatimes.com/videos/city/hyderabad/andhra-couple-commits-suicide-after-recovering-from-covid-19/videoshow/77313492.cms. Accessed on September 11, 2020.

The Wire Staff. (2020). West Bengal: Man who was buying milk dies hours after police Thrash him. *The Wire*, 26 March. Available at: https://thewire.in/rights/west-bengal-police-curfew-man-thrashed-dies. Accessed on Mary 26, 2020.

Umachandran, S. (2020). Blame it on the virus, it's the largest lockdown in history. *Mint*, 26 March. Available at: https://www.livemint.com/news/india/blame-it-on-the-virus-it-s-the-largest-lockdown-in-history-11585159348758.html. Accessed on September 7, 2020.

Chapter 7
Avoidable Disaster Deaths Approach Revisited

Disaster Deaths are Avoidable—
Reducing avoidable disaster deaths is an opportunity for state and non-state actors to redress justice
—Avoidable Deaths Network (on 12 March 2023, ADN, in collaboration with the City of Izumiotsu and the UNDRR—Kobe Office, Launched the Global Campaign 'International Awareness Day for Avoidable Deaths (IAD4AD)'. Thirteen key global messages are launched as part of this Global Campaign, Two of which are quoted here https://iad4ad.avoidable-deaths.net/guidance-for-campaigners/*) (Ray-Bennett and Shiroshita, 2023)*

Abstract This chapter discusses the significance and analytical advantages of the avoidable disaster deaths approach for disaster risk management practice, policy, and research. Most importantly, this chapter identifies disaster risk management and health as the two most important sectors for reducing avoidable disaster deaths. Templates for improving the capacity of some actors in these sectors are presented through the matrices of amenable, preventable and risk governance. The example of Senegal is also discussed to understand the risk governance measures introduced to save lives. The advantages of document review in capturing causes, circumstances, and the number of avoidable disaster deaths are also discussed, along with their scope for replication. Finally, it is recommended that UNDRR considers re-wording the Sendai Target A from 'reducing disaster mortality' to 'reducing avoidable disaster mortality' to usher change in disaster risk management practices and save lives.

Keywords Avoidable disaster death · Disaster risk management · Document review · Sendai Target A

N. S. Ray-Bennett, *Avoiding Disaster Deaths*,
https://doi.org/10.1007/978-3-031-81300-9_7

Introduction

In this book, I have tried to present, analyse, and defend a particular approach that focuses on expanding the capacity of disaster response and health emergency systems. To conclude, I discuss the implications of this approach for practice, policy and research. Although these implications are presented under specific headings, they should be read integrated. This is because they are mutually inclusive. I conclude this chapter by presenting the commonalities and differences between avoidable disaster deaths and the United Nations Office for Disaster Risk Reduction's (UNDRR) disaster death approaches to provide agency to the readers and enthusiasts to experiment in capturing causes, circumstances and number of disaster deaths that are taking place in our global villages, towns, and cities.

Capturing the number of disaster deaths and avoidable disaster deaths need not be national (e.g., Registrar General and Census Commissioner of India and National Crime Records Bureau) and international organisation's responsibility (e.g., CRED's EM-DAT, Sendai Monitor). Instead, as social agents for change, we should take an interest in capturing the causes and circumstances surrounding disaster deaths in our localities. It is important to know 'how and why' a friend, a neighbour or a relative died during a disaster or 'what they died from' so that lessons can be learned, shared, and passed down to the next generation. Also, these questions are essential in guiding decisions in public health so that evidence-based intervention can be developed and implemented to save lives in the next disaster. My methods, matrices, and templates, presented in this chapter's concluding section, should assist with this.

Improve Disaster Risk Management Practice

In Chap. 1, I mentioned that this book aims to help policymakers and practitioners make better decisions to save lives and reduce injuries from disasters by raising awareness about which deaths are avoidable and unavoidable. To do so, I proposed an alternative way of looking into disaster risk management through the lens of avoidable disaster deaths. I redefined UNDRR's definition of a disaster and suggested the following:

> Disaster is a serious disruption of the functioning of primary and secondary disaster responders and services, destruction and disruption of health facilities and services at national and local levels from the impact of a hazard and inadequate disaster risk governance, leading to deaths of ten or more people, 100 or more people are reported affected, injured, and/or homeless and exceeds the ability of the affected community to cope using its resources. (Adapted by the Author based on evidence and from UNDRR and ISC (2020, p. 54), UNISDR (2009))

From the perspective of the complex systems failure approach (Chap. 2), this definition provides a pathway for identifying the relevant professional actors involved in reducing disaster deaths. This is the first step of the avoidable death's perspective (Mansfield, 1970):

- Who is responsible for reducing avoidable disaster deaths?
- Is it an individual, emergency services, medical professional, the local government, or the national government?
- When is the right time to reduce avoidable disaster deaths?
- What are the early warnings?
- What is the treatment?
- How can they be prevented?

This level of clarity is essential in disaster response and recovery. It should promote accountability and develop the capacity of actors and organisations to improve disaster risk management practices.

As observed in Chaps. 2–6, reducing avoidable disaster deaths cross-cuts many sectors and services, including health, disaster risk management, press information bureau, public transport, public food distribution, and emergency services—to mention a few. This finding is consistent with another study by Gupta et al. (2022) in India. These authors identify the interlinkages of various management systems responsible for reducing disaster deaths. This includes (i) medical response and trauma services; (ii) early warning agencies and dissemination; (iii) community-based disaster management, civil defence and community volunteers; (iv) evacuation and transport; and (v) search and rescue. In addition, this study found that the health system, including ambulance and referral services, plays a vital part in reducing amenable and preventable avoidable disaster deaths.

Understanding health systems as integral to disaster risk management systems is nascent in the context of DRR. That said, efforts are underway to raise its visibility and importance through the Sendai Framework's Target A (Substantially reduce disaster mortality by 2030). This is explicit in one of the guiding principles of the Sendai Framework: "Managing the risk of disasters is aimed at protecting persons and their property, health, livelihoods and productive assets, cultural and environmental assets, and the right to development" (UN 2015, paragraph 19(c), p. 13).

Health is currently a key theme, and health resilience is promoted throughout the Sendai Framework's Target D ('Substantially reduce disaster damage to critical infrastructure and disruption of basic services', see Paragraphs 18, 30(i), 30(j), 33(c), and 33(o) (Maini et al., 2017; Ray-Bennett et al., 2022). As such, many of Target's A and B indicators are health-related. For instance, within Targets A-1 (number of deaths and missing persons attributed to disasters per 100,000 population) and B-2 (number of injured or ill people attributed to disasters per 100,000 population) (Alam & Ray-Bennett, 2021; Ray-Bennett et al., 2022; UN, 2015). Within Target D, the number of health facilities damaged or destroyed and the disruption to health services attributed to disasters is to be measured (UN, 2015).

It is also argued that the Sendai Framework's Target A is strengthened through efficient disaster risk governance—the third component of avoidable disaster deaths (Mizutori, 2020; Ray-Bennett et al., 2022). During the first wave of the COVID-19 pandemic, the then UNDRR's Secretary-General, Mizutori (2020), emphasised the importance of saving lives through effective disaster risk governance. Effective disaster risk governance is critical to implementing rights-based disaster risk reduction (Alam & Ray-Bennett, 2021). Although the existing literature is commendable in teasing out the linkages of Target A with Targets B, D and E, and with human rights, health-emergency disaster risk management and disaster risk governance, they are still singular approaches and do not focus fully on capturing direct and indirect disaster deaths from a biological or other disaster (Alam & Ray-Bennett, 2021). Furthermore, the relevant sectors work in silos. I observed this first-hand while researching in India, Bangladesh, and Japan. I found that disaster managers often do not work collaboratively with health systems and emergency actors during or after disasters to reduce disaster deaths. The avoidable disaster deaths approach in this aspect has the advantage of disrupting this theory–practice gap by asking six inter-connected questions mentioned above—reiterated for the readers' convenience.

- Who is responsible for reducing avoidable deaths?
- Is it an individual, emergency services, medical professional, the local government, or the national government?
- When is the right time to reduce avoidable deaths?
- What are the early warnings?
- What is the treatment?
- How can they be prevented?

These six questions are reflected in the definition of 'disaster' presented above.

The UNDRR--Bonn Office monitors disaster deaths through the Sendai Monitor and DisInventar to track them throughout the period allocated to each global target. This system is not designed to reduce avoidable disaster deaths nor to act as a warning signal. It tracks the number of UN Member States keeping records and reports using a designated template. The data received from the UN Member States are reviewed and validated by the UNDRR technical team before being published on the Sendai Monitor and DisInventar. It is posited that the avoidable disaster deaths approach can complement and transform this monitoring system to provide 'warning signals of possible shortcomings' in disaster risk management and emergency systems at national and local levels. However, this would require effective communication, collaboration, and cooperation between UNDRR and the UN Member States through bilateral agreements. This would usher in a new way of managing disaster risk management and health emergency services. This would also promote evaluating disaster deaths as integral to national and state-level disaster risk management policies and programmes. Currently, this is missing (Gupta et al., 2022). Focusing on and collecting data on avoidable disaster deaths at national and local levels would provide a new vantage point from which to improve disaster risk management practices and emergency health systems. Also, it will develop an understanding of other areas and sectors needing priority investments and how to integrate them to reduce deaths and injuries.

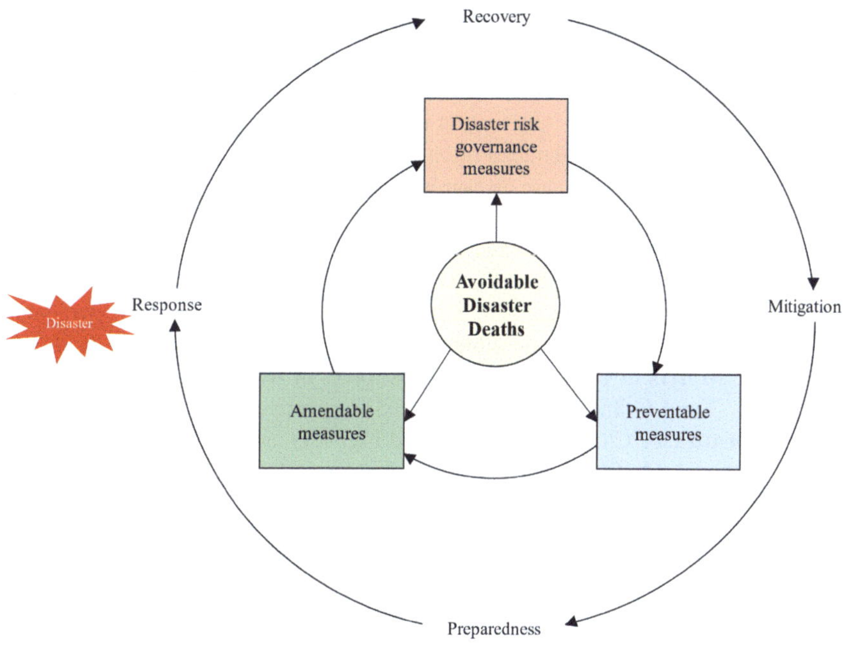

Fig. 7.1 Avoidable disaster deaths approach. *Source* Author

Improve Indicators for the Disaster Risk Management Cycle

Indicators for avoidable disaster deaths at global and national levels are essential milestones or benchmarks (Ray-Bennett, 2018; Wahlström, 2015). They are targets for the international, national, and state disaster management authorities (UN, 2015; Wahlström, 2015). Currently, this is missing from the UNDRR's approach to disaster deaths. In Chap. 3, I explained that the Sendai Framework's 'disaster deaths' are viewed through the lens of 'reducing global disaster mortality by 2030' (Target A). This logic has led to a focus on tracking the target through disaster death stock-taking and monitoring systems (mentioned above), developing indicators to improve methodology and auditing of deaths. As such, the concept of 'disaster deaths' for Target A has not immediately led to a focus on *reducing* these deaths. To overcome this gap, the avoidable disaster deaths approach advocates fewer than ten deaths, which are considered unavoidable in a disaster event (Ray-Bennett, 2018[1]). Furthermore, it is worth considering a country's annual mortality rate as an inevitable unavoidable death to understand excess death during a disaster (Chapter 4). Avoidable disaster deaths are deaths beyond the number ten. They are potentially avoidable from the impact of disasters in the present time through tailored preventable, amenable, and risk governance-related measures (See Fig. 7.1).

[1] This number is inspired by CRED's definition of disaster (Chapter 2).

These measures can be developed and strengthened under the ambit of the disaster risk management cycle of response, recovery, mitigation, and preparedness. Amenable or timely measures should coincide with the response phase. For instance, lifesaving medical intervention is vital to save a life. However, a delay in ambulance service or medical intervention can lessen the chance of survival. Disaster risk governance and preventable measures should be commensurate with preparedness, mitigation, and recovery phases. Otherwise, the response phase will not effectively save lives in a disaster climate. Therefore, the three components of avoidable disaster deaths are the enabler for effective disaster risk management. In this context, it is worth noting that achieving zero damage is impossible, but reducing avoidable disaster deaths is undoubtedly possible. The valuable tools introduced in this book, such as the '**Matrix of Causes and Circumstances of Avoidable Disaster Death**' (see Chaps. 3 and 6, Table 3.3 and Table 6.5, respectively), **Types of Avoidable Disaster Death** (see Chap. 3, Table 3.4, and Fig. 7.1), and 32 broad categories of **Causes of Avoidable Disaster Death** (see Chapter 6, Table 6.2) are few pathways towards reducing avoidable disaster death.

Develop Human and Organisational Capacities

Avoidable deaths in the context of public health and epidemiology are intended only as "warning signals of possible shortcomings" in health care delivery (Charlton et al., 1983, p. 696). Warning signals are helpful for monitoring purposes and eventually lead to reductions in death rates (Charlton et al., 1983). Furthermore, avoidable deaths are intended to signal possible shortcomings in the delivery of health care services as well as disaster management systems. They are not designed to point the finger at poorly serviced health [or disaster areas] (Holland, 1986). Then, how can health emergency and disaster risk management systems be supported in this light because both services play a key role in reducing avoidable disaster deaths? A few practical templates are provided below to reduce avoidable preventable, amenable and risk governance-related disaster deaths.

Amenable Measures

Amenable measures are proposed in Chap. 3 to reduce amenable deaths. Crucial to amenable measures are timeliness and effective healthcare. In Chap. 3, I also introduced a few indicators for checking timeliness and effective healthcare in relation to the emerging literature related to COVID-19 in the year 2020. These indicators, along with three additional indicators based on the findings of this study (availability of hospital beds, access to ambulance services, and referral services), are presented as a **matrix of amenable avoidable disaster deaths** to assist the disaster response system (Table 7.1). All seven indicators demand an increase and improvement in

Table 7.1 Matrix to reduce amenable avoidable disaster deaths

1. The time interval between the onset of symptoms and seeking medical interventions. (e.g., max 5 days for COVID-19 (Rothan & Siddappa, 2020))	2. Time interval between the arrival of the patient and commencing treatment (e.g., max 30 min–1 h for COVID-19 (Baker & Fink, 2020))	3. Active treatment of cases without wasting time on the wrong diagnosis before the correct treatment
4. Percentage of case detection (Fuhrmann & Barbarossa, 2020)	5. Access to ambulance service (max within 40 min from the time of call)	6. Availability of hospital beds. The capacity and availability of beds at district and rural hospitals must be factored in for disaster response
7. Referral service—When a patient is denied service due to a lack of hospital beds, district and rural hospitals should be factored in for referrals for disaster response	8. Availability and accessibility of emergency ambulances. The capacity and availability of ambulances at district, urban and rural hospitals must be factored in for disaster response	

the quality of care and capacity building of health facilities and workforce through skill development, infrastructure, supply of medicines, types of equipment, offering referrals and ambulance services—among other things.

However, seven barriers to amenable measures were also noted based on this study's findings. These barriers have direct relevance to disaster-risk governance measures. For instance, delayed ambulances and low hospital bed availability, delayed ambulances and denial of medical care, and delayed ambulances and denial of medical care (see Chap. 6, Table 6.2). To reduce avoidable amenable disaster deaths, disaster risk and emergency health managers/actors must overcome these barriers. Barriers can be overcome in several ways; in the context of this study, a matrix consisting of eight indicators (see Table 7.1) is proposed based on the reality of the disaster climate of COVID-19. I hope that they provide a pathway to overcome these barriers during disasters. The effectiveness of each matrix indicator can only be achieved in a disaster climate if they are planned and trialled during the mitigation and preparedness phases in readiness for disaster response. Furthermore, relevant practitioners and responders must adjust the time interval, case detection ratio and active treatment time according to a specific hazard[2] disaster.

[2] For instance, a snakebite victim during a flood from a king cobra will require active treatment within 60 min.

Preventable Measures

Public health interventions are vital to avoiding preventable deaths. In Chap. 3, I mentioned that a likely intervention involves *Target* (the entity on which the Action is carried out), *Action* (a deed done by an actor to a Target) and *Means* (the processes and methods by which the Action is carried out) (KSU, 2018). Considering COVID-19, WHO rolled out preventable measures to help curb the spread of the disease (see Table 7.2). In the context of India, economic lockdown and the 'National Disaster Management Act 2005' were used proactively to implement many of these measures advocated by the WHO, including social distancing, quarantine, isolation, wearing of masks, contact tracing, and mass screening. In the context of this study, the causes of preventable death were suicides, murder and poisoning due to consumption of sanitiser. Of these causes—suicides were the highest. Based on some of the stories leading to suicides presented in Chap. 6, it was noted that all the preventable measures advocated by the WHO to the national governments have had a substantial psycho-social impact on individuals at household levels (see the discussion in Chap. 6: Conclusion). These measures created an enabling environment for preventable deaths to occur.

The Ministry of Health and Family Welfare issued a memo on 31 March 2020 entitled 'Minding our minds during COVID-19', which describes the lockdown "can be difficult times for all of us [...]. The most common emotion based by all is Fear. It makes us anxious, panicky and can even possibly make us think, say or do things that we might not consider appropriate under normal circumstances" (MoHFW, 2020a). Although this memo recommends 28 different ways to handle psycho-social aspects, they are predominantly an individual's responsibility to take care of their mind and not of the state government. This practice is unsuitable for a disaster climate, especially in an epidemic or pandemic. Therefore, to reduce preventable deaths, it is recommended that the WHO, national and local governments integrate psycho-social measures as an integral part of public health interventions for disaster response and recovery. The economic lockdown led to social isolation and loneliness among the affected population. Therefore, this additional component must be included on the list of Preventable Measures to reduce suicide or 'other causes of accidental deaths' at home, in hospitals and quarantine centres (Table 7.2).

Disaster Risk Governance Measures

Disaster risk governance is essential to reduce avoidable disaster deaths. In Chap. 5, I mentioned that governance, in general, implies a reduction in the role of the state or the government while concurrently recognising the importance of non-state actors in the governance process. In the first wave of COVID-19, under the remit of the 'National Disaster Management Act 2005', disaster risk governance focused on integrating inter-sector and inter-governmental agencies that discharged services to citizens

Table 7.2 Preventable measures

Means	Target	Action
Social or physical distancing	Keep space between yourself and other people outside of your home	To reduce the chances of contacting COVID-19 through contacts, coughing, sneezing, droplets (to break the chain of transmission)
Quarantine	To keep someone who *might* have been exposed to COVID-19 away from others	To reduce the chances of transmission
Isolate	To separate sick people from healthy people	
Hand hygiene	Regularly and thoroughly clean your hands with an alcohol-based hand rub or wash them with soap and water for at least 20 s	This activity kills viruses that may be on your hands
Respiratory hygiene	When you cough or sneeze, cover your mouth and nose with your bent elbow or tissue. Then, dispose of the used tissue immediately	This activity reduces/prevents the release of droplets into the air and the possibility of someone breathing in droplets, including the COVID-19 virus, if the person coughing has the disease
Wearing of mask	Covering the nose and mouth with a mask	This activity prevents dissemination of virus by positive individuals through nasal and buccal droplets
Contact tracing	Process of monitoring persons who have been exposed to another person infected with a specific disease	To aid in early detection and prompt isolation of new cases
Lockdown	Restricting movement of people either locally or internationally	To reduce chances of transmission as the virus moves from person to person
Mass screening (active case search)	Reaching out to the community and systematically screening the population to find cases of COVID-19	To reduce disease transmission by shortening patients' infectious periods, and earlier diagnosis and treatment improve patients' treatment outcomes
National emergency	A situation in which a government is empowered to perform actions or impose policies that it would typically not be permitted to undertake e.g. instituting a lockdown	To curtail the spread of the virus
Psycho-social care	Reaching out to the community and systematically screening the population to find cases of anxiety, depression, loneliness, etc.	To aid early detection and prompt intervention for counselling or therapy to avoid suicides

Source Adapted by the Author from CDC (2020) and WHO (2020) and findings from this study rather than the involvement of non-state actors (except vaccine manufacturers). The complex and top-down coordination, communication, collaboration, and a combination of these three components of disaster risk governance was noticeable between the Central and State Governments as well as among 16 different Ministries and the Department of Pharmaceuticals, the Directorate General of Foreign Trade, the Department of Science and Technology/Biotechnology and the Indian Council of Medical Research (see Chap. 5).

Although the Central Government-led disaster risk governance successfully curbed the spread of COVID-19 in the first wave (Thayer et al., 2021), it was not conducive to reducing avoidable disaster deaths. As such, disaster risk governance-related deaths were the highest compared to preventable and amenable avoidable deaths (Chap. 6). The significant causes of death were accidental electrocution, exhaustion, exhaustion and starvation, starvation, low hospital bed numbers, medical negligence, motor vehicle accidents, suffocation, and police brutality. Socio-economically vulnerable sections of society mainly carried the burden of these deaths.

On 26 March 2020, The Union Finance and Corporate Affairs Minister announced a Rs. 1.70 lakh crore (~ USD 234,648) relief package for the poor to help them fight the battle against coronavirus (MoHFW, 2020b). The components of the package included: (i) insurance for health workers fighting COVID-19 in government hospitals and health care centres; (ii) Pradhan Mantri Garib Kalyan Anya (food) Yojana[3]; (iii) benefit to 87 million farmers in two instalments; (iv) cash transfers; (v) financial support to women's self-help groups; (vi) directions for the State Governments to utilise the 'Building and Other Construction Workers Welfare Fund' (MoHFW, 2020b). Although this relief package was pro-poor, it was announced relatively late. The stories of the deceased in the first wave of COVID-19 are telling. They indicate that it is essential not only to include food security for the poor and vulnerable during a biological disaster or other types of disasters but also income security and health security (*aka* human security (UNDP, 1994; Ray-Bennett et al., 2010)) early-on as an integral part of disaster risk governance.

An exemplary case in this aspect is Senegal's response to COVID-19 (Sridhar, 2021). The first imported case of COVID-19 was reported on 2 March 2020. The Senegalese government saw the spread of COVID-19 as preventable and took the measures necessary to contain it with as little economic damage as possible. Accordingly, the government responded fast by:

- Setting up an emergency operations centre to lead and coordinate all operations.
- Promoting public health messages to raise awareness about the disease and how to curtail its spread. To improve people's understanding of the pandemic, the government resorted to community mobilisation and dissemination of vital information. Social influencers such as musicians and religious leaders were used to drive its messages to their followers. They used social media channels, radio, television, and flyers to communicate with people in the language they understood.

---

[3] *Aka* the Prime Minister's Scheme for the Welfare of the Poor.

- Providing economic packages to support restrictions, including income and food support to those most vulnerable. To extend support to marginalised groups, the government approved 50 billion CFA francs (US$85 million) to ensure economic support—for food, water and electricity—to the poorest living under restrictions. Senegal received around US$880 million in direct support from donors such as the World Bank, International Monitory Fund and the Islamic Development Bank. The government also allocated 64.4 billion CFA francs (US$115 million) to support the health sector. This included the immediate deployment of health experts, medications, and other necessary medical supplies (Sridhar, 2021, pp. 92–95).

With a population of 16.3 million, with only seven doctors per 100,000 people (compared to 277 per 100,000 people in the United States) and limited resources in everyday life, Senegal rose above all odds (Sridhar, 2021). In *Foreign Policy*'s assessment, Senegal was given 100% for public health directives, 100% for fact-based communication, and 97.2% (second in the world after New Zealand) for the COVID-19 global risk index that assesses the strengths of each country's strategic response in 2020 (Sridhar, 2021). Senegal's response strategy suggests that even in resource-constrained contexts and scientific uncertainty, it is possible to prevent avoidable disaster deaths through effective disaster risk governance.

The components of disaster risk governance coordination, communication, collaboration and 3Cs are identifiable. They can help organisations manage labour, knowledge, and gaps in disaster risk management activities. They can also promote sense-making amongst the organisations by questioning:

- What is 'it' that needs coordinating, communicating and collaborating?
- How can coordination of 'it' be supported, advanced and facilitated by administrative structures, processes and technology?
- What communication techniques are required to enhance coordination between human-to-human, human to technology and technology-to-human interfaces?
- What is the *goal* of this coordination, collaboration and communication?
- Who decides this *goal*?
- How can this goal affect high-risk and vulnerable communities in reducing or increasing avoidable disaster deaths? (adapted from Ray-Bennett, 2018)

The best outcome of a *goal* will be achieved if the decision-making process is based on the rational model, as compared to bureaucratic and political models (Pfeffer, 1981; Ray-Bennett, 2018). Successful outcomes will also be facilitated if the alignment processes of actors and organisations have begun in the mitigation and preparedness phases rather than in a disaster climate. Alignment of actors and organisations for health emergency and disaster response is a prerequisite in a disaster climate. The achievement of a *goal* will also be influenced if the leadership of a response system is context-specific, organisation-specific and hazard-specific. However, an overall *goal* to reduce avoidable disaster deaths will have a far-reaching effect if this *goal* is set by the head of a state or by the head of a nation as compared to a bureaucratic head of a disaster management organisation in the event of a large-scale disaster response (Ray-Bennett, 2018). Unfortunately, the *goal* to reduce avoidable disaster deaths was unclear in India's first wave of COVID-19. No one set a *goal* to reduce avoidable disaster deaths. The *goal* was instead to contain the spread of the virus. Setting a *goal* is paramount to reducing avoidable disaster deaths.

Capacity building of the responding organisations is currently the Sendai Framework's second 'Priority for Action' (Strengthening disaster risk governance to manage disaster risk), which mentions: "empower local authorities as appropriate, through regulatory and financial means to work and coordinate with civil society [...] in disaster risk management at local level" (UN, 2015, p. 13). The three components of avoidable disaster deaths (amenable, preventable and disaster risk governance measures) can significantly contribute towards realising this Priority for Action by developing the capacity of not just local authorities but also of the associated organisations, including the secondary and tertiary.

The three components encapsulate uncertainties, dynamism, imagination, critical decision-making processes, readiness, incident management, coordination, collaboration, communication and combinations of the 3Cs—to mention a few. It is suggested that preparation for avoidable disaster deaths can lead to scenario planning amongst the responders for each component and, in doing so, build the capacity of responders to detect blind spots, human errors and new possibilities (Ray-Bennett, 2018; Senge, 1990; Taleb, 2007). Preparation for each component can also allow responders to expand their imagination and heighten awareness of a crisis (Perrow, 1999; Ray-Bennett & Shiroshita, 2019; Senge, 1990; Weick, 1988). Heightened awareness should enable responders to see the bigger picture and prepare for a disaster. Training, consultation, debrief, and knowledge exchange can all contribute to the capacity building of responders and organisations towards reducing avoidable disaster deaths. Furthermore, planning disaster risk management from the perspective of avoidable disaster deaths can lead to locating gaps and strengths in existing preventable measures, amenable measures and governance-related measures through technologies (related to early warning systems and information and communication tools), as well as in responders' knowledge and expertise. It is posited that this level of planning can lead to new types of research to understand the intersection of public health and epidemiology, disaster risk governance and human security to improve disaster response systems at national and local levels.

Advance Theory and Methods

The three components of avoidable disaster deaths are not without limitations. At times, it was challenging to determine the 32 broad categories of 'causes' of avoidable disaster death under specific headings of preventable, amenable, and governance-related. It was also observed that the 'circumstances' of amenable and preventable deaths directly affect disaster risk governance during the mitigation and preparedness phases of the disaster risk management cycle. Therefore, two new indicators were introduced to reflect the nuanced causes of death: 'unknown deaths' and 'COVID-19 deaths' that can be of any of the three (amenable, preventable and disaster risk governance)'. Therefore, underreporting of specific conditions and/or miscalculation in ascertaining the causes of avoidable disaster death is inevitable. A case study design based on mixed methods research is recommended to rectify this. Gathering first-hand qualitative data (using a social autopsy tool) on preventable, amenable, and disaster risk governance from the perspectives of disaster responders and deceased family members and quantitative data from hospital records can provide a comprehensive understanding of avoidable disaster deaths.

Deaths in general and during disasters, in particular, are likely to occur from a wide variety of causes or multiple factors, including comorbidities, socio-economic factors, access to health care in post-disasters, loss of income, starvation, exhaustion, and medical negligence—among other social determinants of health (Chapter 6). Likewise, ascertaining the circumstances, causes, and manner of avoidable disaster deaths is equally challenging. To overcome some of the challenges in determining the cause of death, the Government of India and WHO recommend postmortem to ascertain the medical cause of death or verbal autopsies where the former is lacking due to a poor vital registration system (Dattani, 2023; also see Chap. 4). Implementing these methods requires resources (funding and research team) and can be time-consuming. Primary data collection using one of these methods was beyond this prospective study's remit.

Disaster deaths are also captured through two epidemiological methods: a special sample survey of deaths and body counts (Guha-Sapir & Chechi, 2018). A special sample survey involves tracking a small part of the population in the absence of complete civil registration[4] (WHO, 2014). The body count approach meticulously compiles records from mortuaries, death certificates and vital registration records and this method has been found effective (Guha-Sapir & Chechi, 2018). However, capturing the number of avoidable disaster deaths in the first wave of COVID-19 using one of these methods was also beyond the remit of this study.

By contrast, using the avoidable disaster deaths approach, I conducted a document review of the emerging grey literature (aided with academic literature) during the lockdown to capture causes, circumstances, and the number of indirect deaths. By doing so, I departed from the abovementioned traditional methods. Although content review of documents has limitations (e.g., information such as causes of death and circumstances surrounding deaths can be missing—as noted during the review), it can be rectified through the triangulation of mixed methods research mentioned above. That said, I hope that the process of document review of grey and academic literature, including newspapers, government portals, reports, and memos (among other sources), excites emerging researchers and scientists from the field of disaster risk reduction, development, sociology, and environment, to take risks to conduct a small-scale prospective study on avoidable disaster deaths. As mentioned above, researching disaster deaths and avoidable disaster deaths need not be the sole responsibility of our national and international organisations. As capable and compassionate individuals, we should all take an interest in exploring 'how and why

[4] In India, one example is the Sample Registration Survey Statistical Report 2018 (ORGI, 2020) for causes of death– discussed in Chapter 4.

people die', 'who dies' and 'who is responsible for saving lives' during disasters so that we can learn lessons and also pass down these lessons to our posterity for growth and sustainability. Furthermore, this data can also act as evidence to promote accountability in the local disaster response and emergency health systems, along with scopes for developing context-specific interventions to save lives.

I introduced 'four categories of actors' that can 'ascertain the causes of avoidable death', a 'matrix of causes and circumstances', 12 variables on how to review the content of secondary data to capture avoidable disaster deaths, a matrix for amenable and preventable deaths and additional components and questions to improve disaster risk governance. These resources are replicable and adjustable. I encourage readers, emerging researchers, and practitioners to apply these resources and expand the area of avoidable disaster deaths in lower-middle and low-income countries.

I introduced seven critical tenets of avoidable disaster deaths in Chap. 3, and I reiterate them for the readers' convenience.

(i) Avoidable disaster deaths may be classified as avoidable and unavoidable. Avoidable disaster deaths are: direct and indirect immediate deaths.
    (a) Direct and indirect avoidable deaths can be: amenable deaths, preventable deaths, governance-related deaths
(ii) Avoidable disaster deaths focus on the causes and circumstances of different types of direct and indirect immediate avoidable disaster deaths
(iii) Avoidable disaster deaths focus on classifying deaths not just for record keeping and data management but also for capacity building and resource distribution for targeted intervention in resource-constrained set-ups
(iv) Avoidable disaster deaths focus on disaggregating different types of direct and indirect deaths and their circumstances to improve policy and practice
(v) Time is important and critical in reducing avoidable disaster deaths
(vi) Avoidable disaster deaths focus on actors and organisations who can increase or reduce disaster deaths
(vii) Avoidable disaster deaths emphasise knowledge on how to reduce immediate and delayed avoidable disaster deaths.

Although these seven tenets differ from the Sendai Framework's disaster deaths approach, they are complementary. This is because both approaches must aim to reduce disaster deaths. In this vein, I have identified a few similarities and differences between these two approaches (see Table 7.3). I recommend that the sector researchers and practitioners use their agency to select the relevant tenets or a combination of them to suit their context to initiate small-scale prospective studies or to improve the disaster response system to reduce avoidable disaster deaths.

Table 7.3 Disaster deaths versus avoidable disaster deaths

Avoidable disaster deaths	Disaster deaths
Classifies deaths as avoidable and unavoidable. Avoidable deaths are indirect disaster deaths	Classifies deaths as direct, indirect, and missing. Only direct deaths and missing persons are recorded
Avoidable disaster deaths can be direct and indirect immediate deaths	Indirect deaths can be delayed and immediate
Direct avoidable deaths can be: amenable deaths, preventable deaths, governance-related deaths Indirect immediate avoidable deaths can be: amenable deaths, preventable deaths, governance-related deaths	Knowledge is missing in this regard
Focuses on the causes and circumstances of different types of direct and indirect immediate avoidable disaster deaths	Focuses on the medical causes of direct deaths
Focuses on classifying deaths not just for record keeping and data management but also for capacity building and resource distribution for targeted intervention in resource-constrained set-ups	Focuses on classifying deaths for record keeping and data management
Focuses on disaggregating different types of direct and indirect deaths and their circumstances to improve policy and practice	Knowledge is missing in this regard
Time is important and discussed	Time is important but not discussed for indirect deaths
Actors who can increase or decrease avoidable disaster deaths are identified	Actors who can increase or decrease disaster deaths are not identified
Knowledge on how to reduce immediate and delayed avoidable disaster deaths is offered	Knowledge of how to reduce immediate and delayed deaths is missing

Source Adapted by the Author from IRDR (2015) and Ray-Bennett and Shiroshita (2023) and findings from this study

Conclusion

The Sendai Framework's Target A to 'reduce disaster mortality by 2030' is a step forward. Still, it is essential to understand that the actual burden of disaster deaths or mortality can only be reduced when causes, circumstances, barriers, and enablers are understood at local and national levels. Equally, it is essential to improve the knowledge of disaster responders and practitioners as to which deaths are avoidable and unavoidable in a disaster climate. I hope this book's avoidable disaster deaths approach offers this critical understanding.

Last, I recommend that UNDRR consider re-wording the Sendai Framework's Target A from 'reducing disaster mortality' to 'reducing avoidable disaster mortality'. This should usher *change* in disaster management practices and save many precious lives within seven years of the Sendai Framework's timeline of 2030 and beyond.

References

Alam, E., & Ray-Bennett, N. S. (2021). Disaster risk governance for district-level landslide risk management in Bangladesh. *International Journal of Disaster Risk Reduction, 59*(2021), 102220. https://doi.org/10.1016/j.ijdrr.2021.102220

Baker, M., & Fink, S. (2020). It's just everywhere already: How delays in testing set back the U.S. coronavirus response. *New York Times*, 11 March. Available at: https://www.nytimes.com/2020/03/10/us/coronavirus-testing-delays.html. Accessed on March 11, 2020.

CDC (Centers for Disease Control and Prevention). (2020). *Social distancing, quarantine, and isolation.* Available at: https://www.cdc.gov/coronavirus/2019-ncov/prevent-getting-sick/social-distancing.html. Accessed on April 21, 2020.

Charlton, J. R. H., Silver, R., Hartley, R. M., & Holland, W. W. (1983). Geographical variation in mortality from conditions amenable to medical intervention in England and Wales. *The Lancet, 321*(8326), 691–696. https://doi.org/10.1016/s0140-6736(83)91981-5

Dattani, S. (2023). How are causes of death registered around the world? *OurWorldinData.org*. Available at: https://ourworldindata.org/how-are-causes-of-death-registered-around-the-world. Accessed on October 27, 2024.

Fuhrmann, J., & Barbarossa, M. V. (2020). The significance of case detection ratios for predictions on the outcome of an epidemic—A message from mathematical modelers. *Archives of Public Health, 78*(63), 2–6.

Guha-Sapir, D., & Chechi, F. (2018). *Science and politics of disaster death tolls—Science must prevail.* Centre for Research on the Epidemiology of Disasters. Available at: https://www.cred.be/science-and-politics-disaster-death-tolls-science-must-prevail. Accessed on July 19, 2020.

Gupta, A. K., Chopde, S., Bindal, M. K., & Baidya, S. (2022). *Institutional mechanism for evaluating human deaths from disasters.* Researchgate. Available at: https://www.researchgate.net/publication/362347714_Institutional_mechanism_for_evaluating_human_deaths_from_disasters. Accessed on April 4, 2023.

Holland, W. (1986). The 'avoidable death' guide to Europe. *Health Policy, 6*, 115–117. Available at: https://www.sciencedirect.com/science/article/abs/pii/0168851086900011. Accessed on October 28, 2020.

IRDR (Integrated Research on Disaster Risk). (2015). *Guidelines on measuring losses from disasters: Human and economic impact indicators (IRDR DATA Publication No. 2).* IRDR. Available at: http://www.irdrinternational.org/wpcontent/uploads/2015/03/DATA-Project-Report-No.-2-WEB-7MB.pdf. Accessed on July 10, 2020.

KSU (Kent State University). (2018). *A look at public health interventions.* Available at: https://onlinedegrees.kent.edu/college-of-public-health/community/public-health-interventions. Accessed on April 21, 2020.

Maini, R., Clarke, L., Blanchard, K., & Murray, V. (2017). The Sendai framework for disaster risk reduction and its indicators—Where does health fit in? *International Journal of Disaster Risk Science, 8*(2), 150–155.

Mansfield, J. (1970). *Avoidable deaths.* Sussex, Littlehampton Book Services Ltd.

Mizutori, M. (2020). Taking stock of disaster losses five years after the Sendai world conference. *The Japan Times*, 17 March. Available at: https://www.japantimes.co.jp/opinion/2020/03/17/commentary/japan-commentary/taking-stock-disaster-losses-five-years-sendai-world-conference/?fbclid=iwar1rjvpxwhaancfvmsgyssd1xd9fj43ovckodyrb314s1oyfytfdwjjcbva#.XnIWC6j7SUl. Accessed on July 18, 2020.

MoHFW (Ministry of Health and Family Welfare). (2020a). *Minding our minds during the COVID-19.* Ministry of Health and Family Welfare. Available at: https://www.mohfw.gov.in/pdf/MindingourmindsduringCoronaeditedat.pdf. Accessed on March 17, 2021.

MoHFW (Ministry of Health and Family Welfare). (2020b). *Finance Minister announces Rs 1.70 Lakh Crore relief package under Pradhan Mantri Garib Kalyan Yojana for the poor to help them fight the battle against Corona Virus.* Available at: https://www.mohfw.gov.in/pdf/MoFPMGaribKalyanYojanaPackage.pdf. Accessed on March 17, 2021.

Perrow, C. (1999). *Normal accidents: Living with high-risk technologies* (2nd ed.). Princeton.

Pfeffer, J. (1981). *Power in organizations*. Pitman Publishing.

Ray-Bennett, N. S. (2018). *Avoidable deaths: A systems failure approach to disaster risk management*. Springer Nature.

Ray-Bennett, N. S., Clarke, K., & Mendez, D. (2022). Sendai framework's global Targets A and B: Opinions from the global platform for disaster risk reduction's ignite stage 2019. *International Journal of Disaster Risk Science*, 2022(13). https://doi.org/10.1007/s13753-022-00432-3

Ray-Bennett, N. S., Collins, A., Bhuiya, A., Edgeworth, R., Nahar, P., & Alamgir, F. (2010). Exploring the meaning of health security for disaster resilience through people's perspectives in Bangladesh. *Health and Place, 16*(3), 581–589. https://doi.org/10.1016/j.healthplace.2010.01.003

Ray-Bennett, N. S., & Shiroshita, H. (2019). Disasters, deaths and the Sendai framework's target one. *Disaster Prevention and Management: An International Journal, 28*(6), 764–785. https://doi.org/10.1108/dpm-09-2019-0302

Ray-Bennett, N. S., & Shiroshita, H. (2023). *International Awareness Day for Avoidable Deaths (IAD4AD): Guidance for campaigners*. Avoidable Deaths Network. Available at: https://iad4ad.avoidable-deaths.net/wp-content/uploads/2023/02/IAD4AD_Guidance_for_Campaigners.pdf. Accessed on February 28, 2023.

Rothan, H., & Siddappa, H. (2020). The epidemiology and pathogenesis of coronavirus disease (COVID-19) outbreak. Available at https://www.researchgate.net/publication/339515532_The_epidemiology_and_pathogenesis_of_coronavirus_disease_COVID-19_outbreak. Accessed on April 20, 2020.

Senge, P. M. (1990). *The leader's new work: Building learning organizations*. MIT Sloan Management Review. Available at: https://sloanreview.mit.edu/article/the-leaders-new-work-building-learning-organizations/

Sridhar, D. L. (2021). *Preventable: How a pandemic changed the world & how to stop the next one*. Viking.

Taleb, N. N. (2007). *The black swan: The impact of the highly improbable*. Penguin Books Ltd.

Thayer, M., Hasan, Md. Z., Sankhla, P., & Gupta, S. (2021). An interrupted time series analysis of the lockdown policies in India: A national-level analysis of COVID-19 incidence. *Health Policy and Planning, 36*, 620–629. Available at:https://doi.org/10.1093/heapol/czab027

UN (United Nations). (2015). *Sendai framework for disaster risk reduction 2015–2030*. UNDRR. Available at: http://www.unisdr.org/we/inform/publications/43291. Accessed October 9, 2024.

UNDP (United Nations Development Programme). (1994). *Human development report 1994*. Available at: https://hdr.undp.org/content/human-development-report-1994. Accessed on October 22, 2020.

UNDRR (United Nations Office for Disaster Risk Reduction) and ISC (International Science Congress). (2020). *Hazard definition and classificaiton review: Technical report. A report developed by Sendai framework for DRR 2015–2030*. International Science Congress (ISC) and United Office for Disaster Risk Reduction, Geneva. Available online https://www.undrr.org/publication/hazard-definition-and-classification-review-technical-report. Accessed on September 20, 2023.

UNISDR (United Nations International Strategy for Disaster Reduction). (2009). *2009 UNISDR terminology on disaster risk reduction*. UNISDR. http://www.preventionweb.net/files/7817_UNISDRTerminologyEnglish.pdf. Accessed on February 11, 2025.

Wahlström, M. (2015). *From agreement to action: What next for the new global framework on disasters?* The Policy Lab Meeting at the Royal Society.

Weick, K. E. (1988). Enacted sensemaking in crisis situations. *Journal of Management Studies, 25*(4), 305–317. https://doi.org/10.1111/j.1467-6486.1988.tb00039.x

WHO (World Health Organization). (2014). Civil registration: Why counting births and deaths is important. Available online https://www.who.int/news-room/fact-sheets/detail/civil-registration-why-counting-births-and-deaths-is-important. Accessed on December 29, 2023.

WHO (World Health Organization). (2020). *WHO director-general's opening remarks at the media briefing on COVID-19 - 11 March 2020*. [Online] Who.int. Available at: https://www.who.int/director-general/speeches/detail/who-director-general-s-opening-remarks-at-the-media-briefing-on-covid-19---11-march-2020. Accessed on April 20, 2020.

Index

N. S. Ray-Bennett, *Avoiding Disaster Deaths*,
https://doi.org/10.1007/978-3-031-81300-9

E

The manufacturer's authorised representative in the EU is Springer Nature Customer Service Centre GmbH, Europaplatz 3, 69115 Heidelberg, Germany. If you have any concerns regarding our products, please contact ProductSafety@springernature.com

Printed and bound by CPI Group (UK) Ltd, Croydon, CR0 4YY
01/10/2026
02225944-0011